▼ Hua Luogeng's lecture manuscript for young students in the 1950s

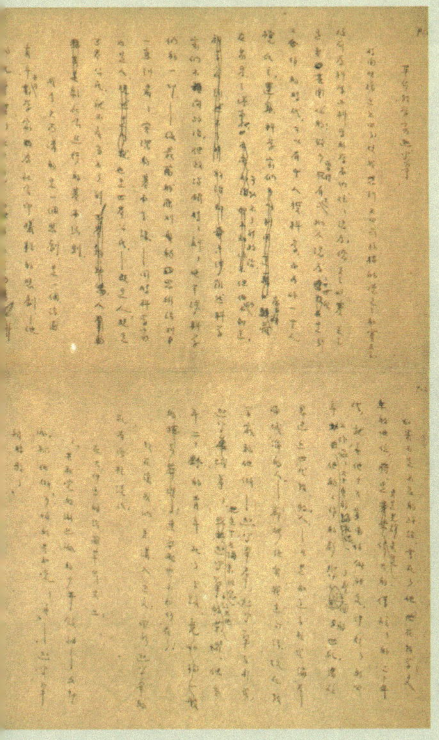

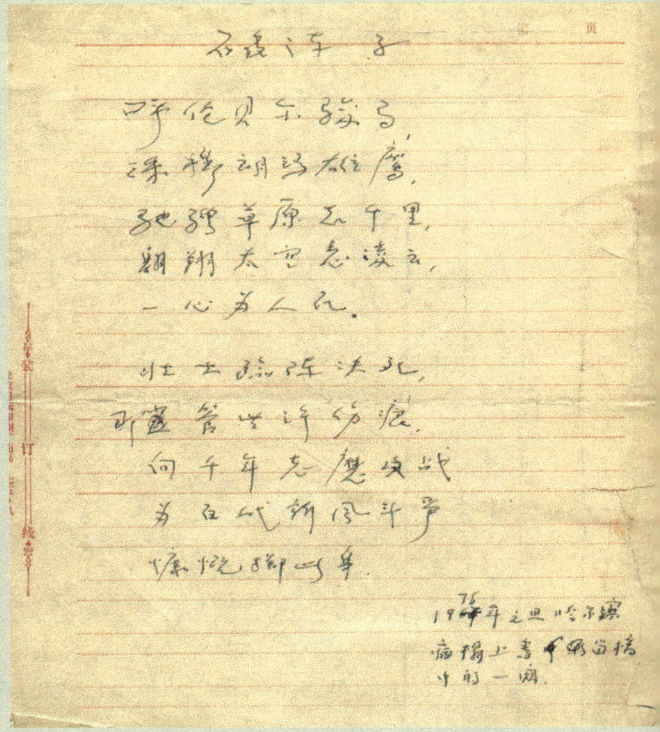

破阵子

呼伦贝尔骏马，
珠穆朗玛雄鹰，
驰骋草原五千里，
翱翔太空志凌云，
一心为人民。

壮士临阵决死，
哪管些许伤痕，
向千年老魔作战
为百代新风斗争
慷慨掷此身。

▲ Hua Luogeng's poetry draft "Pozhenzi"

► Hua Luogeng at Cambridge University (1936)

▲ Hua Luogeng (back row, second from the left) on the ship returning to China (1950)

▲ Hua Luogeng discussing mathematical problems with researchers in front of the Mathematics Institute at Tsinghua Garden (1953)

▲ Hua Luogeng giving a lecture to students of the Class of 1958 at the University of Science and Technology of China (1958)

▼ A gathering of masters (from left to right: Hua Luogeng, Laoshe, Liang Sicheng, and Mei Lanfang) (1954)

◀ Hua Luogeng with oil workers in Daqing (1972)

◀ In May 1979, Hua Luogeng was approved to join the Communist Party of China. This was his first time attending a party branch meeting, where he was warmly welcomed by fellow members.

▲ Nancy University (now the University of Lorraine) in France awarded Hua Luogeng an honorary doctorate (1979).

▲ Hua Luogeng signing at the US National Academy of Sciences conference (1983)

◀ Hua Luogeng's last known photograph (1985)

▲ Hua Luogeng revisiting Princeton University (1980)

▲ Hua Luogeng performing mathematical calculations on a fan

▲ Hua Luogeng visiting Birmingham, UK

▲ Despite his advanced age, Hua Luogeng remained tirelessly dedicated to mathematical research.

◀ Hua Luogeng in his later years

Hua Luogeng

Innovating Independently in Mathematics

www.royalcollins.com

Innovating Independently in Mathematics

Library *of* Mathematicians' Thoughts

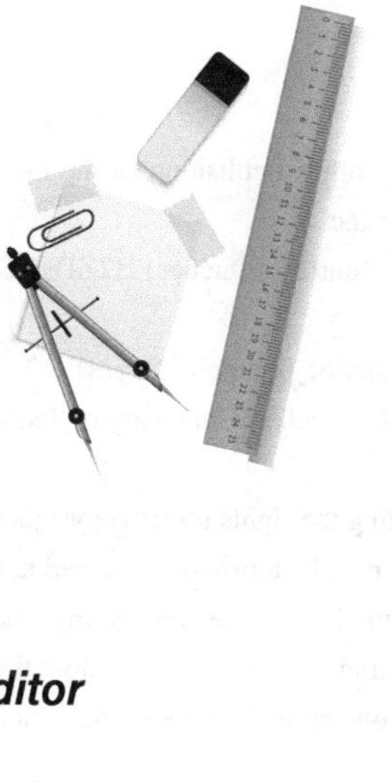

Collected Edition Editor ***Li Wenlin***

Author ***Hua Luogeng***

Editors ***Li Wenlin***

Books Beyond Boundaries

ROYAL COLLINS

Innovating Independently in Mathematics
Library of Mathematicians' Thoughts

Collected Edition Editor: Li Wenlin
Author: Hua Luogeng
Editor: Li Wenlin

First published in 2025 by Royal Collins Publishing Group Inc.
Groupe Publication Royal Collins Inc.
550-555 boul. René-Lévesque O Montréal (Québec) H2Z1B1 Canada

ISBN: 978-1-4878-1360-4

Contents

Preface to the Collected Edition

The first volume of *Library of Mathematicians' Thoughts* was released in 2009, followed by the completion of the second volume in 2021. Now, the publisher has decided to present a beautifully bound collected edition, bringing together two volumes into one exquisite compilation. Ten prominent modern mathematicians come together to share their insights on the essence, meaning, and value of mathematics, as well as to pass on the methods and spirit of mathematical innovation ... The masters' thoughts are presented in their purest form.

The purpose and significance of the *Library of Mathematicians' Thoughts* has been elaborated in detail in the preface to the first and second volumes, "Reading the Masters, Approaching Mathematics," so I will refrain from repeating it here.

At this moment, China is moving toward the fulfillment of its second centennial goal. In this era of innovation-driven rejuvenation of the Chinese nation, the promotion and dissemination of scientific culture and the enhancement of the scientific literacy of the Chinese people carry immense strategic importance. We sincerely hope that the publication of this collected edition of the *Library of Mathematicians' Thoughts* will continue to shine a light on the modern endeavor of spreading mathematical culture and fostering the spirit of science.

Along with the necessary textual revisions, this collected edition includes photographs capturing the activities of the featured mathematicians. Some chapters also feature supplementary appendixes that enhance readability, infusing the serious nature of this mathematical library with a lively and engaging touch.

From the original publication of the first volume to the present collected edition, over a decade has passed. During this time, the dedication of the editors and translators, the perseverance of the publisher, and the invaluable support and feedback from our readers have been indispensable. As I stand before the proofs of the ten-volume collected edition now ready for release, I share in the satisfaction and reflection of the editors, while also feeling that our work is far from complete. We will continue to forge ahead!

Li Wenlin

November 2022

Zhongguancun, Beijing

Reading the Masters, Approaching Mathematics

■ Mathematical Thought: The Soul of Mathematicians

Mathematical thought is the soul of mathematicians. Imagine this: Without axiomatic thinking, could we ever have Euclid or Hilbert? Without the fusion of numbers and geometry, where would Descartes be? Without the concept of mathematical structure, how could we even begin to speak of the Bourbaki school?

The essence of a mathematician's thought is primarily embodied in their groundbreaking mathematical research—through the new concepts, theories, and methods they introduce. Newton and Leibniz's calculus, Gauss, Riemann, and Lobachevsky's non-Euclidean geometries, Galois' notion of "groups," Gödel's incompleteness theorems and Turing machines, Nash equilibrium theory, and so on—all come together to form an awe-inspiring ocean of mathematical thought, shaping the indelible chapters of human intellectual history.

But the mathematical philosophy of these mathematician is also part of mathematical thought. Their reflections on the nature, feature, significance, and value of mathematics, their views on the sources of mathematical knowledge, and its connection with other fields of human inquiry, as well as their insights into scientific methodology—these too are integral components of their mathematical thought. Of course, throughout history, mathematicians have often held differing, sometimes opposing views on these issues. Yet, it is precisely these diverse voices that converge to create a symphony of rational thought.

Just as one comes to understand and appreciate an artist or a composer through their paintings or compositions, mathematical thought is undoubtedly the primary way in which we come to know and assess mathematicians. It is also the core of their contribution to humanity and the very substance from which people seek knowledge. In this light, we might say: "Mathematicians think, and thus mathematicians exist."

■ The Societal Significance of Mathematical Thought

Is mathematical thought something only mathematicians should possess? Certainly not. Mathematics is the foundation of natural sciences, technology, and the humanities—a fact that has become increasingly recognized as a cornerstone of modern society. This foundational role of mathematics is not only due to its use as the language and tool of science, finding applications in nearly every field of human knowledge, but perhaps even more so for its profound cultural function: to nurture and develop the human mind, especially its capacity for precise thinking. Regardless of the career one pursues, one's thinking ability is an invaluable, invisible asset, and mathematics is the perfect discipline to hone this ability. This is why mathematics remains one of the subjects that demands the most time throughout a person's education.

This, however, does not mean that every specific mathematical concept learned in school will have direct, practical use later in life. The true impact of mathematics on lifelong development lies in its transformative power over one's way of thinking. Take Euclidean geometry, for example. While most of the theorems we study in school may rarely find direct application in our daily lives or professional work, the rigorous deductive reasoning and logical thinking inherent in Euclidean geometry play an undeniable role in shaping the elite minds of every field. In fact, from Newton's *Mathematical Principles of Natural Philosophy* to Einstein's writings on relativity, from the Declaration of the Rights of Man and of the Citizen during the French Revolution to Marx's *Capital*, and even in the works of modern Nobel laureates in Economics, one can trace the influence of Euclidean thought. Moreover, the quantitative thinking embedded in mathematics has, with remarkable breadth and depth, permeated nearly all areas of human knowledge. From its rigorous proofs to its precise calculations, mathematics offers humanity a model of exact and refined thinking.

A striking example can be found in the concept of "program memory" or the idea of "program automation," which plays a pivotal role in modern computer design. When the first electronic computer, ENIAC, was initially constructed, it faced the imminent risk of failure due to the sharp conflict between its increasing computational speed and the sluggishness of manual programming. At this critical juncture, it was the brilliant mind of mathematician John von Neumann who introduced the concept of "program memory," ultimately rescuing this groundbreaking technological invention. Even today, the fundamental principles of computer architecture continue to adhere to von Neumann's visionary ideas. For this reason, he is rightly revered as the "father of the computer" (although it is now recognized that he was not the sole mathematician to propose such an idea). Isn't it fascinating that a concept like "program memory," which at first glance might not seem "mathematical," was created by a mathematician—by a genius mathematician like von Neumann?

Thus, we can assert that the mathematical thought of mathematicians is a priceless treasure for society as a whole. The propagation of mathematics, beyond the transmission of specific mathematical knowledge, lies at the heart of spreading mathematical thought itself. In today's world, where science and technology are increasingly shaped by mathematics, this has become an urgent societal necessity. Just imagine: If more and more individuals could harness mathematical thinking to think and solve problems, how profoundly that could shape the progress of society!

■ Reading the Masters, Approaching Mathematics

Mathematics is the art of numbers and shapes, and the creative minds of mathematicians are as vibrant as they are original. Their thinking is neither bound by convention nor can it be reduced to rigid dogma. Understanding a mathematician's thought can be approached in many ways, but reading the original works of mathematicians—especially the great masters—is perhaps the most direct, reliable, and rewarding path.

Mathematicians' writings can generally be categorized into two types. The majority is their technical monographs, where they discuss their theories and methods. For those engaged in truly original research, these foundational works are the most vivid and instructive of texts. As Laplace often advised his students: "Read Euler, read him—he is the teacher of us all." Here, Laplace was speaking primarily to professional mathematicians and physicists; for the general public, such works are often difficult to approach.

The other category of works is more accessible, often written for a broader audience. These include reflections on the philosophy of mathematics—what Godfrey Harold Hardy referred to as "mathematics about mathematics"—as well as more popular expositions of mathematical knowledge and the mathematician's own creative contributions. Unlike the obscure technical monographs, these writings are highly readable, more easily embraced by the public. Many of these works are not only intellectually enriching but also artistically elegant. Indeed, several great mathematicians were also masters of language, and if writing is considered an art of language, then these works exemplify the union of mathematics and art. Reading these masterpieces is an aesthetic experience, offering a way to learn about mathematics while savoring the beauty of the language, immersing oneself in the culture and allure of mathematics. This is the very purpose of compiling *Library of Mathematicians' Thoughts*.

Library of Mathematicians' Thoughts selects representative works by some of the most renowned modern mathematicians, with one chapter dedicated to each mathematician. The translation went one after another, and was compiled into two volumes for publication. The first volume offers a curated collection of writings from 20th-century mathematical masters such as D. Hilbert (1862–1943), G. Hardy (1877–1947), J. von Neumann (1903–1957), Bourbaki (1935–present), M. F. Atiyah (1929–2019), etc. (the latter three being reprints). It has been warmly received and reprinted multiple times. Encouraged by this success, we continued with the second volume, which features the writings of mathematicians such as F. Klein (1849–1925), H. Weyl (1885–1955), A. N. Kolmogorov (1903–1987), Hua Luogeng (1910–1985), and Shiing-Shen Chern (1911–2004). These works, often short but brimming with profound insight, have circulated widely, both in China and abroad. They are, in essence, like precious gems in the vast ocean of mathematical thought or beautiful blossoms in the garden of mathematics.

We do not expect these gems or blossoms to reverse the tide of utilitarianism, but we believe in the words of Einstein, who, in honoring Newton, said: "**The products of understanding outlast the noise of passing generations. They endure for centuries, continuing to radiate light and heat**."

We sincerely hope that the "products of understanding" of the great mathematicians featured in this collected edition will continue to shine brightly, spreading the light of mathematical thought and advancing the cause of scientific culture.

Reading the masters and approaching mathematics—everyone stands to benefit.

Li Wenlin

Researcher, Institute of Mathematics and Systems Science

Chinese Academy of Sciences

July 2021

Zhongguancun, Beijing

Preface

This year marks the 105th anniversary of Professor Hua Luogeng's birth and the 30th anniversary of his passing. On this special occasion, we fondly remember him, cherishing his kindness and guidance.

Professor Hua was a great mathematician, whose outstanding contributions to multiple fields of mathematics are widely recognized. As J. W. Richard once noted: "His work in mathematics education and research, both at the secondary and university levels, along with his efforts to popularize mathematics among workers and farmers, is just as significant as his contributions to pure mathematics."

Perhaps our understanding of this aspect of his legacy is not yet fully aligned. Furthermore, Professor Hua was a self-taught mathematician who rose to greatness despite having only a junior high school education. International literature has even compared him to the renowned self-taught Indian mathematician S. I. Ramanujan, as noted by scholars such as N. Wiener and P. Bateman. This suggests that, beyond studying Professor Hua's mathematical achievements, we should also focus on learning from his approach to scholarship, his teaching methods, and his spirit of innovation. In particular, we should draw lessons from the hardships he overcame throughout his life.

Professor Wenlin has carefully curated this volume, *Selected Works of Hua Luogeng*, as part of a distinguished series on renowned mathematicians' academic writings. This collection features 22 carefully selected essays from Professor Hua's popular science works, focusing on three key areas:

1. experiences and methods for learning mathematics
2. patriotic sentiments
3. popular science writings on applied mathematics

With the exception of a few technical articles in the third section, this collection is accessible to both high school-level readers and young professionals.

One particularly notable essay in the third section, "Finite and Infinite, Discrete and Continuous," is a general report on the process, insights, and achievements of Professor Hua and his students in exploring, studying, and researching applied mathematics. However, it requires a solid understanding of calculus and linear algebra, making it more suitable for readers with at least a first-year university mathematics background.

A true highlight of this collection is the first-ever publication of Professor Hua's historic speech, "Innovating Independently in Mathematics." This was his inaugural address in 1952 upon becoming the director of the Institute of Mathematics at the Chinese Academy of Sciences. In this speech, he outlined three major research directions for the institute: pure mathematics, applied mathematics, and computational mathematics—an idea he had actually proposed to the Nationalist government as early as the 1940s. More importantly, he emphasized the significance of innovation and talent development, along with concrete measures to achieve these goals. Wenlin discovered this valuable speech in archival documents.

Although decades have passed since Professor Hua wrote these essays, their insights remain timeless and continue to offer significant reference value.

Professor Wenlin has made outstanding contributions to the field of mathematical history, establishing himself as a pioneer and leading scholar in the study of modern Chinese mathematics. I, too, have a deep passion for this field and have often sought his expertise. His unique perspective in curating this collection adds immeasurable value to the study of Hua Luogeng's legacy.

I extend my sincere gratitude to Professor Wenlin for allowing me to contribute to this compilation, giving me the opportunity to express my deep admiration and heartfelt remembrance of Professor Hua. Last, I wish this volume great success.

Wang Yuan

October 12, 2014

Foreword

The name Hua Luogeng is one that resonates deeply across China. Over the years, numerous collections of his writings have been compiled and published. This particular volume, *Innovating Independently in Mathematics*, takes its title from one of his most significant speeches—a report delivered on the eve of the official establishment of the Institute of Mathematics at the Chinese Academy of Sciences. In this address, Hua boldly set forth a strategic vision: innovating independently in mathematics.

As a pioneer of modern Chinese mathematics, Hua devoted his life to securing China's rightful place on the global mathematical stage. That he articulated such an ambitious goal as early as 1952 speaks volumes about his extraordinary foresight and unwavering dedication. His vision remains deeply inspiring, compelling us to admire his remarkable ability to look far beyond his time. In this report—essentially his inaugural speech—Hua systematically analyzed the historical and contemporary state of Chinese mathematics, laid out the research directions for the newly established institute, and proposed concrete strategies for cultivating talent. Even today, in an era that places a strong emphasis on independent innovation, his insights remain profoundly relevant. Notably, this speech has never been formally published before—this book marks its first official appearance.

The collection opens with a letter Hua wrote in 1950 while journeying back to China from the United States, addressed to all Chinese students studying in the US. Brimming with profound patriotism and heartfelt devotion, the letter continues to inspire strong nationalist sentiments to this day.

Hua's journey as a self-taught scholar is nothing short of extraordinary. With only a junior high school diploma, he ascended to the ranks of world-class mathematicians, earning membership in multiple national academies of sciences. His story serves as a beacon of inspiration for young people striving to scale the peaks of scientific discovery. A significant portion of this book is dedicated to his reflections on how to study and conduct mathematical research, featuring some of his most renowned insights: "Intelligence lies in learning, genius comes from accumulation,"

“From thin to thick, and from thick to thin,” and “Do show off your skills before an expert.” These timeless words of wisdom, distilled from his own experiences, continue to guide countless aspiring scholars along the path of scientific innovation and discovery.

While Hua is best known internationally for his pioneering contributions to analytic number theory, algebra, and complex function theory, his perspective on mathematics as a whole was broad and profound. This is clearly reflected in the “Innovating Independently in Mathematics” speech and in another article featured in this collection, “The Great Utility of Mathematics,” which offers keen insights into the interplay between mathematical theories and applications. This piece has since become a widely cited classic in discussions on the practical applications of mathematics.

Beyond theoretical work, Hua was also a great practitioner of applied mathematics. He was among the first to apply number theory methods to numerical analysis, leading to what became internationally recognized as the “Hua-Wang Method.” Beyond academia, he was deeply committed to bringing mathematics into everyday life, traveling tirelessly across China to popularize mathematical methods among workers and farmers. This was an unprecedented feat in the history of mathematics, both in China and worldwide. To showcase this aspect of his career, this book includes works such as “Finite and Infinite, Discrete and Continuous” and “Popular Explanation of Optimization and Its Supplement,” which illustrate Hua’s unparalleled contributions to applied mathematics and scientific outreach.

In addition to these writings, the collection contains several other important works by Hua Luogeng. However, one especially noteworthy highlight is the first-ever publication of another previously unreleased manuscript: “Revolutionary Mathematician Galois.” Written in the early 1950s, this biographical essay on Évariste Galois was discovered among Hua’s handwritten drafts, which were generously donated by his family to the Institute of Mathematics and Systems Science at the Chinese Academy of Sciences. Together with “Innovating Independently in Mathematics,” which remains unknown before as well, this newly unearthed work distinguishes this collection from previous anthologies of Hua Luogeng’s writings.

I would like to extend my heartfelt gratitude to Academician Wang Yuan for his unwavering support in compiling this volume, to Hua’s family for their generosity in donating his manuscripts, and to Dalian University of Technology Press for their dedication in bringing this book to life. Their efforts—especially in deciphering, organizing, and digitizing Hua's aged, faded, and often difficult-to-read handwritten drafts—are deeply appreciated.

Li Wenlin

Institute of Mathematics and Systems Science
Chinese Academy of Sciences
July 2018
Zhongguancun, Beijing

An Open Letter to All Chinese Students Studying in the United States*

Dear friends,

As I bid you farewell and return to our homeland ahead of you, I find myself overwhelmed with emotions. Though words fail to fully express my thoughts, I can assure you that every sentence in this letter is filled with sincerity, spoken from the depths of my heart.

Frankly speaking, what I am about to share is the result of an intense inner struggle over the past year. The reasons behind my decision to return to China have been shaped both by personal reflection and discussions and correspondence with friends. Friends, if you have ever felt a similar turmoil, I hope this letter will serve as a reference for your own decisions. If you have not yet encountered such thoughts, I urge you to read on, for the emergence of such feelings is far from coincidental.

Let us begin with the broader picture. The world today is clearly divided into two camps: one that strives for the well-being of the people and another that exists solely for the interests of a privileged ruling class. The former stands on the side of justice, guided by truth; the latter is riddled with contradictions. One aligns itself with oppressed nations, while the other builds its so-called "civilization" upon the suffering of the unfortunate. Every citizen of the world must make a choice: For the sake of humanity, we must stand with truth and justice, and align ourselves with the interests of the majority.

Friends, let us reflect deeply. We have experienced immigration restrictions and racial discrimination—each a barrier imposed upon us. It is true that some so-called "exceptional" individuals have managed to break through these barriers and have

* Written on the journey back to China in February 1950.

been granted the "privilege" of naturalization. But if we think with a clear conscience, how can we take pride in our own acceptance while our fellow countrymen continue to face oppression and discrimination? To rejoice in personal recognition while turning a blind eye to the plight of our people—what kind of heart allows this? And let us be honest: those who have been "accepted" are merely being used for others' purposes.

Some may argue that democracy and freedom are values to be cherished in this society. But I urge you, do not be deceived by mere words. The illusion of these ideals has misled people for far too long.

Consider the reality: The capitalists control all the means of communication—radio, newspapers, magazines, and films. Their voices dominate the world, while ours are but a whisper against the din of their drums and gongs. Yes, we may all have "freedom," but I ask you: When one's hands grow weary and the strings of the instrument snap, will anyone still hear the melody of the solitary lute? To speak of "democracy" in a society of economic inequality is self-deception; to speak of "freedom" in such a world is to willingly accept shackles. True freedom and democracy can only exist within true equality. Without equality, these so-called ideals are nothing more than tools of the ruling class.

Let us take a moment to reflect carefully: How did we come to study abroad? Perhaps we believe it was purely through our intelligence and hard work that we passed the exams and earned the opportunity to go overseas. Perhaps we think it is our abilities and skills that have enabled us to establish ourselves here. Consequently, one might conclude that the privileges we enjoy are the fruits of our own talents, and that our standing here is the result of our own efforts. However, from my perspective, this conclusion does not tell the whole story. Why is that? Who provided the special opportunities that allowed us to complete our university education? Who supplied the necessary foreign currency that enabled us to study abroad? Was it not our hardworking compatriots? Was it not our parents who toiled and sacrificed for us? Having received the support of our people and the sweat and blood of our families, how can we now turn away and refuse to serve them? How can such an attitude be called fair or just? Friends, we must not betray those who have nurtured us. We must see that since we have received the privilege of education, we must fulfill our corresponding duty. Those of us who are gifted and capable bear an even greater responsibility—to dedicate ourselves to the enormous task of building our great new nation!

Now, let us turn our attention to our newly reborn homeland, advancing steadily upon the foundation of a great victory! The New Year's Message from Xinhua News Agency states as follows:

> 1949 was the year in which the Chinese People's War of Liberation achieved a great victory and the People's Republic of China was proclaimed. That year, we shattered the so-called "peace offensive" of domestic and foreign reactionaries, wiped out the Kuomintang bandits across the Chinese mainland, and liberated over 90 percent of the population, securing a fundamental victory in the war. That year, representatives of the democratic forces nationwide convened the Chinese People's Political Consultative Conference, which adopted the Common Program as the fundamental law of the country and established the Central People's Government. This government not only gained the overwhelming support of the Chinese people but also received widespread recognition from the anti-imperialist forces worldwide. The Soviet Union and other People's Democracies quickly established equal and friendly diplomatic relations with China. That year, we liberated and took control of major cities and vast rural areas, swiftly establishing revolutionary order, suppressing counterrevolutionary activities, and organizing the laboring masses. In many cities, people's representative assemblies were convened, while in rural areas, bandits were eradicated, a fair taxation policy was implemented, and campaigns of rent and interest deduction and campaigns against landlords and feudal oppression were launched. Despite enemy sabotage, blockades, and severe droughts and floods, we overcame these difficulties, restored production and transportation as much as possible, and achieved significant progress despite financial imbalances …
>
> China is advancing rapidly. The victories of 1949 far exceeded expectations, both in scale and speed. In 1950, with conditions even more favorable than in the previous year, our achievements will surely surpass our current expectations at a faster speed. Once armed resistance is eradicated across China and as the people's awareness and organization continue to strengthen, our country will gradually overcome the severe difficulties left by years of war and steadily embark on the path to prosperity.

Friends! "Though the foreign land is beautiful, it is not our home"—let us go home.

Some may say, "I am still young; there is no harm in staying here a little longer." But I say, "Not necessarily." We are in the prime of our lives. If we are destined to return, why delay? Why not return now and devote our energy where it is most needed?

In conclusion, we must return for the sake of truth. We must return for our nation and our people. We must return to serve our people. Even for our own futures, we must return—to lay the foundations of our careers and contribute to the construction and development of our great motherland!

Friends, with these earnest words, let us reunite this year in our capital, Beijing!

Mathematics: What Chinese Excel At*

In the past, imperialists not only exploited us economically and politically, transforming our country into a semi-colonial and semi-feudal state, but also waged a cultural assault. Through their churches, schools, hospitals, and so-called charitable institutions, they sought to undermine our national dignity and self-confidence. Political oppression is overt, often bloody; economic exploitation is palpable, causing immense suffering. But cultural invasion, though subtle, begins with a seemingly sweet exterior and ends with us forgetting our ancestors, accepting the oppressors as our rightful rulers. The brilliance of this strategy lies in its quiet infiltration of our consciousness, making us believe that we are inferior in every way, that we fall short of others. Such poison can erode our spirit and determination, leading us to a place of irreversible despair.

Yet, the great people of our nation have made unparalleled contributions to human history. Even in fields where some mistakenly believe we are "lacking," such as certain sciences, our accomplishments have proven that we are not as deficient as others would have us believe. To enumerate all of our achievements would be impossible in a short article, and is beyond my own knowledge. Here, I will provide just a few examples from mathematics, a field I know something about. I encourage you to approach these examples with an objective mindset and a fair heart, and judge for yourselves whether we are truly the "inferior race" that imperialists claim, or whether the self-doubt that some harbor, or the poison already seeping through their thoughts, is merely the result of false belief that "we are bad at science."

Before we begin, I must clarify that I am not a historian of Chinese mathematics, nor do I claim to have exhaustive knowledge. This article's purpose is simply to remind my fellow countrymen that mathematics is indeed one of our strengths.

---

* Originally published on February 10, 1951, in *People's Daily*.

Whether the exact dates of these discoveries are fully confirmed, or whether the examples provided are comprehensive, is not the focus here. I do not intend to boast about our nation's superiority or to claim that we are inherently better than others. I believe both superiority complexes and inferiority complexes are distortions of reality. Only imperialists make distinctions between the "superiority" and "inferiority" of races, using such theories as a basis for exploitation and oppression. Yes, those who make discoveries early should be celebrated, but the fact that some nations may have discovered things later does not make them inferior. Culture is a reflection of economics and politics, and judging a nation's intelligence based solely on the timing of their inventions would be unfair and biased.

■ Pythagorean Theorem

An interesting thought experiment was once proposed: If there were intelligent beings on other planets, and we wanted to communicate with them, how could we do so? It is clear that neither written language nor spoken words would work. Even pictures would lose their meaning, because their forms of life might be completely different from ours. What we consider "human form" might be their "monstrous appearance," and customs might differ as well. Our "salute" could be their "battle stance." In this case, someone suggested using mathematical diagrams as a universal medium of communication. While this is a playful thought, it underscores an important point: Mathematics is a universal truth. The mathematical diagram in question was developed by our ancient ancestors and has been recorded for over two thousand years! It is, of course, a creation of the laboring people, originally used for defining right angles, calculating areas, and measuring heights. Its origins long predate written records. Our ancient texts not only mention this particular principle but also go further to describe: "Multiply the lengths of the legs of a right triangle individually, add them together to form the hypotenuse, and then take the square root to determine the length of the hypotenuse." In modern language, this means: "The sum of the squares of the two legs of a right triangle equals the square of the hypotenuse (fig. 1)." This is known in the West as the Pythagorean theorem, yet the Chinese formulation of this principle predates Pythagoras by centuries.

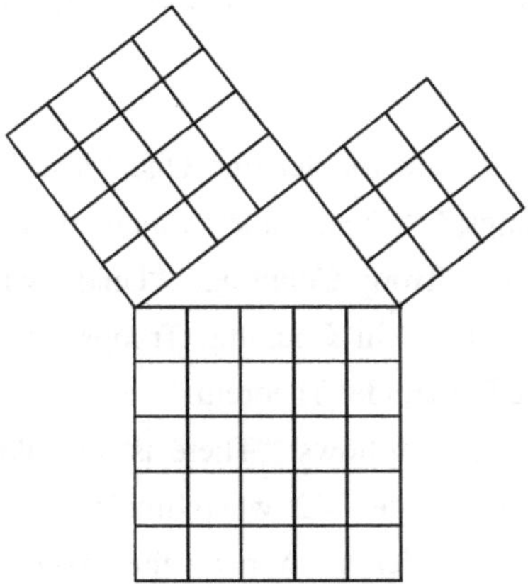

Figure 1

Pi (π)

When it comes to pi, we also have a glorious history. The earliest records of the approximation of pi, where the ratio of the circumference to the diameter is roughly 3, date back to ancient times. Liu Hui's "circle-squaring" method from the Wei-Jin period (around 263) not only laid the foundation for calculating pi, but also clarified the fundamental principles of integral calculus, particularly in calculating lengths and areas. He used polygonal approximations, gradually increasing the number of sides to approach the curve. Starting with a hexagon, he proceeded to a 12-sided, 24-sided polygon, and so on, progressively dividing the circle to calculate pi. Liu Hui arrived at an approximation of pi as 3.1416.

Later, during the Southern Dynasties (420–589), Zu Chongzhi (429–500) achieved even greater precision, predicting the emergence of asymptotic values. He demonstrated that pi lies between 3.1415926 and 3.1415927. Furthermore, he used the fractions $\frac{22}{7}$ and $\frac{335}{113}$ as approximation and close ratio respectively, which, in modern terms, are known as "best approximations." These two fractions are the first terms in the series of continued fractions, with the next term being highly complex. Zu Chongzhi's approximation predates that of the German mathematician Otto by over a thousand years (Otto's record was in 1573).

Great Derivation Method for Seeking One

The "Great Derivation Method for Seeking One," also known by various names such as "Counting Things," "Ghost Valley Calculation," "Wall-Adjacent Calculation," "King Qin's Secret Troop Counting," "Unknown Total of Things," "Pipe Cutting Method," and "Han Xin Counting Troops," is referred to by Western scholars as the "Chinese Remainder Theorem."

The problem is stated as follows: "There is an unknown number of things. When divided by 3, the remainder is 2; when divided by 5, the remainder is 3; and when divided by 7, the remainder is 2. What is the unknown number?"

The algorithm is accompanied by a mnemonic rhyme: "Three people traveling, the remainder is seventy. Five trees with twenty-one flowers, seven children gathered at the fifteenth. Divide by one hundred and five to get the answer."

The method is as follows: Multiply the remainder of the division by 3 by 70, the remainder of 5 by 21, and the remainder of 7 by 15. Add these products together, then subtract the nearest multiple of 105 to get the answer. For example, in the given problem, the calculation is $2 \times 70 + 3 \times 21 + 2 \times 15 = 233$. Subtract 210 (the nearest multiple of 105), and the result is 23.

This method not only holds an esteemed position in history but, when compared with foreign methods in number theory, it remains superior. It is concrete, simple, and easy to apply.

The origin of this method can be traced back to the *Sunzi Suanjing* (*The Mathematical Classic of Sunzi*), which predates Greek mathematician Diophantus. The method was further developed by Qin Jiushao in his *Shushu Jiuzhang* (*Mathematical Treatise in Nine Sections*) in 1247, which was about 500 years earlier than European mathematicians such as Euler (1707–1783), Lagrange (1736–1813), and Gauss (1777–1855). Additionally, Qin Jiushao also invented the Euclidean algorithm.

Yang Hui's Triangle

The triangular pattern (fig. 2) follows a construction rule where both sides are equal to one. Each number in the triangle is the sum of the two numbers above it. This triangle forms the fundamental algorithm of the binomial theorem. It is commonly referred to in the West as Pascal's Triangle (named after Blaise Pascal, 1654). However, according to mathematical historians in the West, the first person to invent this method was Apianus (1527). In contrast, Chinese mathematicians such as Yang Hui (1261), Zhu Shijie (1303), and Wu Xinmin (1450) discovered this concept over 200 years earlier, predating Apianus by over two centuries.

1
1 1
1 2 1
1 3 3 1
1 4 6 4 1
1 5 10 10 5 1
1 6 15 20 15 6 1

Figure 2

Qin Jiushao's Polynomial Solutions

The method of solving numerical equations known as Horner's method is a fundamental technique in algebra. It was invented by William George Horner in 1819. However, upon examining the history of Chinese mathematics, it is clear that the underlying principles of this method were already known in *Yigu Genyuan* (*Discussions on the Origins of Antiquity*) around 1080. Over time, this method was further developed by Liu Yi and Jia Xian, and by 1247, Qin Jiushao had a complete version of the method, predating Horner by 572 years. Subsequent mathematicians, such as Li Ye (1248) and Zhu Shijie (1299), also applied this method, continuing its use more than 500 years before Horner.

(In ancient times, astronomy and mathematics were closely intertwined. Our contributions to astronomy, such as Guo Shoujing's work on precession, are also significant but are beyond the scope of this article.)

Certainly, if we continue to dig deeper, we will discover more, better, and more valuable materials. It is important to acknowledge, however, that after the late Yuan Dynasty (1206–1368), Chinese mathematics experienced a period of stagnation and even regression. This was not due to a decline in intellectual ability, but rather the result of shifting political and social environments. The militaristic focus of the Yuan Dynasty, along with the rigid examination systems of the Ming Dynasty (1368–1644), hindered scientific progress. Meanwhile, the production situation in China remained stuck in the feudal society phase, while Europe, after the Renaissance, transitioned into a capitalist society. As a result, their mathematics advanced rapidly, leading to the significant differences we see today.

Yet this gap is temporary, not intrinsic.

The interpretation of these words is not difficult. In ancient times, we entered the stage of civilization earlier (as described by Engels), so the development of our mathematics began ahead of Europe. While Europe was still in its dark ages, we had already made remarkable contributions. We should take pride in and draw inspiration from the great achievements of our ancestors, but we must not become complacent or fall into the trap of superiority. The rise of European capitalism (which, at the time, carried progressive elements) accelerated the development of mathematics, whereas we temporarily lagged behind. However, this temporary setback should not lead us to feel inferior or believe that we are falling short in every aspect. When we compare capitalist societies with our new democratic, socialist, and communist societies today, we see that the difference between them spans an entire era. Hence, I firmly believe that, in the near future, when the cultural renaissance foreseen by Chairman Mao arrives, our mathematics—indeed, our entire science and culture—will experience a remarkable leap forward and occupy a significant place on the global stage.

Innovating Independently in Mathematics*: A Report on the Goals, Directions, and Tasks of the Institute of Mathematics, Chinese Academy of Sciences

In the summary report of this session, Director Guo has officially announced that our institute has been established. Along with its establishment comes the honorable responsibility of serving the people. The crucial question now is how we can execute our work both correctly and effectively while also aligning with the needs of national development. This is the key topic we need to explore. I will begin by presenting a few preliminary thoughts as the foundation for our discussion. While we must give due attention to the immediate and specific issues at hand, we must not lose sight of the broader, long-term goals. We need to focus not only on the depth of our work but also on its foundational aspects. While it's important to concentrate on our current state, we must also be aware of the larger context outside the Institute of Mathematics. In sum, we should approach our goals from a long-term perspective, then integrate them with the present environment to devise a present, step-by-step, and feasible work plan.

* This article is the speech draft prepared by Director Hua Luogeng prior to the official establishment of the Institute of Mathematics at the Chinese Academy of Sciences in 1952. The speech was later refined and supplemented (see appendix) before being submitted for approval to the academy's director and deputy director. It was subsequently delivered at a full assembly of the institute. The title of this article has been added by the editor.

Mathematics has long been a field in which our ancestors excelled, and in the past three decades, it has remained one of the most accomplished sciences. In recent years, the number of papers authored by Chinese mathematicians has hovered around a hundred annually. Considering the relatively short history of modern mathematics, these achievements are certainly commendable. Furthermore, in some specialized departments, our contributions have gained significant recognition within the global mathematical community, including fields such as differential geometry, Fourier analysis, and fluid mechanics, etc. Reflecting on our past, rich with remarkable achievements, and the excellent environment fostered by our new democratic society ready to prosper, we are filled with boundless confidence. Our future is incredibly bright; we will certainly fulfill our noble mission and joyfully embrace the cultural renaissance that Chairman Mao has predicted.

Although our past accomplishments are worthy of praise, the dark rule of the Chiang regime created undeniable shortcomings.

The semi-colonial society we endured cast a long shadow over our mathematical research. For example, the problems we addressed were often sourced entirely from foreign journals, and some methods were simply adopted from abroad. Once completed, our work was sent to foreign journals, and if published, we were pleased with our success. In academic institutions and research bodies, the senior staff were often those who had studied abroad and returned home. This imbalance placed our scientific endeavors in a subordinate position. Of course, this was a natural consequence of the old society, where the Chiang regime did not promote the publication of scientific journals. As a result, papers were often sent overseas. And the Chiang regime was subservient to the United States, and naturally, some scientists took pride in receiving approval from Americans. Our research shows that since the May Fourth Movement, approximately one thousand papers were written, yet fewer than one-tenth were published within China.

Moreover, in the past, there was a notable lack of comprehensive planning. A prime example of this is the narrow-minded approach taken by the Institute of Mathematics of the "Academia Sinica." They not only excluded applied mathematics from their research scope, but also limited their focus to a very narrow portion of theoretical mathematics. From the perspective of individual heroism, it may seem effective for a leader to focus all their energy on a narrow field, as it can quickly yield visible results and allow the leader to maintain tight control over the work. However, when considering the broader interests of the people and the long-term development of science, this approach is fundamentally flawed. The people require comprehensive, holistic science, not fragmented, isolated branches. Our future cultural renaissance will be independent, all-encompassing, and grand, rather than merely a token gesture toward American imperialism or a superficial ornament adorning foreign cultures (though perhaps I am overly flattering!).

In the past, collective work was rarely attempted, and to this day, we have yet to establish consistent, organized, and focused discussions on specific topics. This can primarily be attributed to the old society's disinterest in promoting such practices, compounded by the ingrained habit of "scholars looking down on one another." Without collective discussions, there has been little opportunity to exchange experiences, learn from one another, or engage in collaborative work. As a result, the work we do is often understood only by those directly involved, leading to a situation of self-congratulation, detached from the masses and disconnected from reality. This, in turn, fosters a mentality of "factionalism" or "my way or no way."

Reflecting on these past shortcomings, the guiding principles for our institute's work are beginning to take shape. In order to correct the "semi-colonial" research practices of the past, we must establish an autonomous, national research approach. The conditions for this are now fully in place. The new democratic society we are building is exceptionally conducive to the implementation of planned, collective scientific research. Furthermore, Chinese mathematics is entering such a new phase. In other words, since we began absorbing Western mathematical methods, we have clearly passed through several stages: from learning to imitation to "localized creativity." We are now in a phase where the number of individuals engaging in creative work is rapidly increasing. Therefore, I am confident that the objective conditions for fostering independent mathematical research are now fully mature.

However, I do not mean to suggest that we should cease learning from and imitating the outstanding works of foreign mathematicians—particularly from Soviet scholars. On the contrary, we should actively draw from a broad range of sources to strengthen our foundation and accelerate our growth. But this absorption must not be mindless; it must be active, thoughtful, and critical. Only by doing so can modern mathematics flourish on our own soil. To emphasize once again, the conditions for mathematical research are now robust, and with our dedicated efforts, we will undoubtedly take a leading role in the forthcoming cultural renaissance.

Science is not a collection of isolated pearls to be admired by the idle elite; it is a complex machine made up of interlocking gears—each one serving the people. If any single gear fails to function, the entire machine is affected, either reducing its efficiency or causing it to grind to a halt. Therefore, the development of science must be holistic, with each discipline playing its part. Just as each gear is interdependent on its neighboring gears, each scientific field is interconnected with others—both in theoretical foundations and in practical applications. If we fail to recognize and nurture these connections, we will fail to unlock the full potential of science and miss out on understanding its evolution.

At this stage, our institute's work is focused on both theory and application. Each researcher should specialize in their area of expertise but also remain aware of adjacent fields. However, just as gears come in different sizes, we cannot afford to treat every gear the same. While no gear should be overlooked, we must recognize that they each serve a distinct purpose. We should carefully consider factors such as the scope, urgency, complexity, and future potential of each task when allocating resources. Meanwhile, we must take into account the current conditions and resources at our disposal. Only by doing so can we ensure that our plans are implemented effectively and efficiently.

In terms of foundational theory, we are in a much stronger position than before. While our leadership capabilities are still growing, we are prepared for broad, comprehensive development. At present, we aim to ensure that key areas like algebra, geometry, and analysis have researchers dedicated to them. Looking further ahead, as we specialize further, we hope that each subfield will have experts, and that in the further future, every important area of mathematical theory will have specialists here in China.

As for applied mathematics, there are still considerable gaps across the country, and our institute has largely concentrated on gas dynamics. Thus, we must convene planning meetings to assess the national situation, particularly the distribution of talent across different fields. This will then guide our future plans. However, we must also be mindful that applied mathematics is the area most likely to overlap with other departments, so while identifying gaps is essential, we must also avoid unnecessary duplication.

In China, computational mathematics has long been neglected. Yet, its significance in the scientific landscape is undeniable, as it provides essential computational support for various fields that require extensive calculations. To support the advancement of other scientific disciplines, we must commit to fostering and expanding this discipline. Our vision is that within three to five years, we will have the necessary infrastructure for computational mathematics, including acquiring the requisite machinery and training skilled personnel who can both operate and understand its structure.

In China, collective research is still a relatively new attempt, but we must recognize its purpose: to amplify, not diminish, our collective strength. In the past, the scholarly culture of mutual disparagement weakened our progress. At the same time, superficial harmony, devoid of meaningful discourse, is equally detrimental. In essence, collective research means having a group of dedicated individuals working together in a structured and well-coordinated manner, rather than engaging in aimless discussions that waste time and energy. To ensure effective collaboration, we must maintain close cooperation with the Chinese Mathematical Society and maintain constant communication with the specialized committees of our institute. If preparations permit, we plan to organize one or two specialized academic conferences this year, bringing together experts from across the country to exchange knowledge and insights on specific mathematical problems.

At this stage, in addition to the tasks mentioned above, our main areas of focus also include the following departments:

We must conduct a comprehensive nationwide survey of the mathematical community. This goes beyond collecting basic information such as names, ages, and places of origin. Instead, we seek to understand each individual's areas of research, expertise, and academic background. Our goal is to compile an index of mathematical papers and conduct a thorough investigation of existing mathematical literature in China. With this information, and in alignment with patriotic principles, we will follow Comrade Hu Qiaomu's directive to disseminate this knowledge to the public. We will assist the Compilation and Translation Bureau in the challenging yet crucial task of curating a collection of selected research papers. Additionally, standardizing mathematical terminology is an urgent need in China, and we are committed to supporting this effort.

The study of advanced experience will undoubtedly play a decisive role in the future of China. To acquire the necessary tools, our institute will launch a Russian language learning initiative this year, aiming for all our colleagues to develop the ability to read Russian mathematical literature within a year. Only by mastering this essential tool can the slogan "Learning from the Soviet Union" become more than just empty words.

To support the comprehensive development of our research fellows, I propose the following academic expectations: Within two years, every member should achieve proficiency in "Three Advanced" areas—Advanced Algebra, Advanced Geometry, and Advanced Calculus—while also developing a specialized expertise in one particular field. To strengthen your grasp of these Advanced areas, our institute will offer a seminar on linear group theory next year, which will further solidify and enhance your knowledge of algebra and geometry. Additionally, Comrade Wu Xinmou's course on differential equations will supplement your analytical skills, serving as an essential tool for applied mathematics. As for developing a specialized expertise, my expectations are even more ardent, as this will lay the foundation for comprehensive future advancements. However, given the limited number of experienced researchers available to lead such studies, we must adopt a focused and steady approach. This year, our primary research priorities at the Beijing institute are: … In addition, we will continue collaborative work with affiliated researchers at Zhejiang University, Peking University, and Tsinghua University.

Comrades, please do not forget that while you are on the path to becoming specialists, you are first and foremost mathematicians. Therefore, always keep in mind the importance of a well-rounded development in mathematics and remain aware of its interconnected fields. Likewise, remember that you are members of this research institute, and as such, you should take ownership of its progress by actively offering suggestions and constructive criticism. Most importantly, never forget that you are citizens of People's Republic of China. So, do not confine yourselves to a narrow professional bubble, indulging in self-satisfaction.

To become good citizens of People's Republic of China and outstanding cadres of Chairman Mao, we must continuously study Marxism-Leninism and Mao Zedong Thought, while also keeping up with current affairs and policies—this is essential for every citizen in a new democratic society. Some may ask, "Why should a scientist need to understand politics?" Others may argue, "Natural science is classless, and studying it will not turn us into counter-revolutionaries." However, such views distort Stalin's directive. He stated, "Natural science is not class-based," but he also made it clear that "in a class society, science is usually controlled by the ruling class." Take, for example, a nuclear physicist—he can either develop murder weapons in the United States or contribute nuclear energy for the benefit of the people in China. The same applies to mathematicians: They can use computational techniques to serve the capitalist killing machine, or they can return to People's

Republic of China to educate engineers for the nation's progress. Developing this fundamental understanding requires dedicated and consistent study. Only through study can we truly understand ourselves, our country, and the world.

Regarding the attitude toward learning, Chairman Mao once said: "The enemy of learning is self-satisfaction. To truly acquire knowledge, one must begin with humility—'never tire of learning' for oneself and 'never tire of teaching' others. This is the attitude we should adopt."

Indeed! This approach should apply not only to political learning but also to professional study. Stalin also said: "The higher the political level and Marxist-Leninist understanding of personnel in any field of state and party work, the better and more effective their work will be, and the greater their achievements."

Therefore, the foundation of all our future work will undoubtedly be based on learning.

At this stage, I hope that everyone will actively participate in administrative tasks. Since this work benefits us all, it is our collective responsibility to undertake it. In particular, efforts in research, academic paper compilations, and terminology standardization must move forward most swiftly. At the same time, the development of mathematics across the nation is also a matter of our concern. Therefore, we must maintain close ties with the people, assist them in solving problems, and provide principled guidance.

We must not forget the damage inflicted upon our institute by the Chiang regime. They looted all our books and shipped them to Taiwan region. Our staff were scattered in an effort to escape tyranny, forcing us to rebuild everything from scratch. In particular, our access to books remains a significant challenge due to the blockade imposed by American imperialists.

However, neither the oppression of the Chiang clique nor the obstruction of American imperialism can hinder our progress. Under the wise leadership of Chairman Mao and with the support of the Communist Party of China, we have both the courage and confidence to fulfill our mission. We are determined to restore the prosperity of mathematics—the great pillar of the natural sciences—in China. We will reclaim the honored place our ancestors once held in the history of mathematics. We will stand as a powerful force in the new cultural renaissance foreseen by Chairman Mao.

Comrades! For the motherland! For the people! Let us march forward with unwavering resolve!

Appendix: Supplements and Revisions[1]

Guiding Principles

We must establish a pragmatic and pioneering work ethic—one that does not pursue abstraction for its own sake or popularization merely for the sake of popularization. Abstraction and popularization must always be grounded in substantive content, avoiding aimless theorization.

The original draft mentioned learning from the Soviet Union, but it lacked sufficient persuasion and emphasis. We intend to underscore the importance and urgency of learning from the Soviet experience. (To this end, studying and reading Russian will not only be a regular task this year but will also include an intensive training session.)

In alignment with the overall development of mathematics in China, fields with existing achievements will continue to advance, while those with gaps will be systematically supplemented. Our aim is to follow these developmental directions.

Developmental Directions and Phases

The long-term vision set by the Soviet Union serves as our current goal: "The Institute of Mathematics plans to develop all major branches of modern mathematics. In addition, special attention will be given to differential equations in mathematical physics, the application of probability theory in technology, and the compilation of various specialized mathematical tables."

---

1 The Institute of Mathematics of the Chinese Academy of Sciences has provided the following explanation when submitting the Supplement and Revision to the Academy's Planning Bureau: "Dear colleagues at the Planning Bureau, the notice issued by the General Office on September 12 outlined the directives for our responsible colleagues regarding the current status of our institute (including existing issues), as well as the guidelines, tasks, development direction, and steps. A report was to be prepared within our institute, with the report outline submitted to the Planning Bureau for further review. The Planning Bureau, in collaboration with our institute's responsible colleagues, will address the matter together, and once the content is confirmed, it will be submitted for approval to the director and deputy director before proceeding with the official report. In accordance with these instructions, we are presenting the speech initially drafted by Director Hua Luogeng, prepared prior to the establishment of our institute after the 1951 Academy Administrative Meeting. We intend to proceed with this speech, incorporating the attached 'Supplements and Revisions,' for the upcoming report."

Considering both the national context and the realities of our institute, our direction and steps are as follows:

1. Fundamental mathematics. First, gradually expand research across all major areas of modern mathematics. At this stage, our key focus areas include differential equations, several complex variables, analytic number theory, topology, and probability theory. Our affiliated researchers, working at their respective institutions, focus on Kähler space differential geometry (Su Buqing), univalent function theory (Chen Jiangong), representation theory of groups (Duan Xuefu), quantum mechanics (Zhang Zongsui), and foundations of mathematics (Hu Shihua). Second, priority recruitment areas for talent: functional analysis and integral equations.
2. Applied mathematics. First, our objective is to progressively expand research topics with direct industrial applications and connections to related scientific fields such as physics and engineering sciences. Second, given the interdisciplinary nature of this institute, collaboration with relevant institutions is essential to avoid research gaps and redundant efforts. Third, we will select and send scholars to study in the Soviet Union, gaining advanced knowledge and addressing research gaps.
3. Computational mathematics. This is a crucial yet underdeveloped field in China that must be advanced. At present, we are in the preparatory stage.

▸ Missions

The key points regarding the "missions" are outlined for three specific groups:

1. Research interns. A structured training plan has been developed. The original 1951 draft's principle of "mastering the Three Advanced and specializing in one field" requires further clarification: First, among the Three Advanced, the emphasis is on analysis (the term Three Advanced will be phased out in the coming years). Second, since mathematics students in the past have largely neglected mechanics, additional studies in this area will be supplemented. Third, specialization in one field will be pursued in the second and third years, with the primary task of research interns being learning—focusing on mutual assistance

and collective progress.

2. Researchers and associate researchers. They should establish their own research plans and take full responsibility for mentoring research interns. Guiding research work is a fundamental duty, which should be conducted through structured seminars rather than informal or unregulated study sessions.
3. Assistant researchers. They should formulate their own research plans and take on specific responsibilities in training research interns.

These points serve as guiding principles rather than rigid regulations. For instance, if a research intern develops an original research plan that receives approval from the leadership, they should be encouraged to proceed.

Revolutionary Mathematician Galois*

As the time for the May Fourth Movement approaches, I am reminded of the ideals of Democracy and Science advocated during that period. In the past, scholars who focused purely on science often believed that Democracy and Science were unrelated—at least some thought that the twentieth century, being an era of division and cooperation, would allow some to engage in scientific pursuits while others worked on democratic movements. Scientists were supposed to remain isolated in their ivory towers, disconnected from politics. However, they failed to recognize that while scientists may ignore politics, politics continuously interferes with every aspect of a scientist's life—from the most abstract principles to the most basic daily routines. Scientists, too, are human beings and global citizens, and as such, they cannot remain ignorant of the fundamental laws governing human society.

What I wish to speak of today is a tragedy, the story of a young French mathematical genius who sacrificed his life for society—his life taken at the age of only twenty-one.

Had it not been for harmful political forces, his position in the history of mathematics would have been even more glorious and monumental. In just a brief span of twenty years—if we consider that he began his research at sixteen—his four short years of work and just sixty pages of papers made an impact that surpassed the entire century. Those of you familiar with modern algebra—those who understand the theories of equations, group theory, and field theory—will be able to verify my

* This manuscript is a report delivered by Hua Luogeng to young students in the early 1950s, though the exact year remains to be confirmed. It was never formally published during Hua Luogeng's lifetime. The manuscript was discovered by the editors of this book among the collection of Hua Luogeng's personal papers, which were donated by his family to the Institute of Mathematics and Systems Science at the Chinese Academy of Sciences. With the assistance of Dalian University of Technology Press, the manuscript has been carefully organized, transcribed, and is now being published for the first time.

words. Today, mathematicians commemorate him with terms such as Galois group, Galois equation, and Galois field to honor his contributions. Yet, after this brilliant young man died in his 20s, even his burial place was unknown.

Now, let us turn to the core of the matter—there are two explanations for Galois's death:

In the history of the French Revolution as narrated by Dumas, "The young leader of the radical Republican Party—whom the reactionaries called a terrible Republican—was assassinated."

In the history of mathematics, "A brilliant comet streaked across the summer night sky. This young genius mathematician died in a duel over a love affair."

The two different accounts present contrasting images of Galois: According to the first, he is a brave and fearless warrior, his character as wild as the sea's stormy waves and the thunder in the sky. In the second account, he appears to be a charming, passionate, and sensitive intellectual, his nature like the spring in southern China, longing to live out his life in the gentle embrace of love. According to many stubborn individuals, mathematicians need introverted personalities, able to sit at their desks for hours, lost in thought, while revolutionaries require extroverted personalities, those who shout and charge into battle. But can these two types of characters exist in one person? History proves that it is indeed possible! As long as one loves truth, one can excel as a scientist in the realm of science and as a dedicated revolutionary in the realm of politics. This is not a contradiction, but a unity of two facets of reality.

Galois was born on October 25, 1811, in the small town of Bourg-la-Reine, near Paris. His grandfather was the only founder of a primary school in the town that was not under the control of monks. His father was a leader of the liberal movement and, after Napoleon's escape from Elba, was elected mayor of Bourg-la-Reine. He was always among the townspeople, fighting against the monks. Galois was homeschooled by his mother until the age of eleven, and at the age of 12, he moved to Paris to attend Louis-le-Grand School in 1823. The school building was said to resemble a prison, but the courageous memories of the revolution were still fresh, and within them, there simmered a fiery revolutionary sentiment. Without a doubt, Galois's first lesson in revolution began here.

Initially, Galois was a good student, but gradually, he became less concerned with formal subjects. When he reached the penultimate year of school, the principal suggested to his father that it might be best for him to repeat the year, as his health was poor and he had not yet matured. Though he was physically weak, the principal's comment about his immaturity was not entirely accurate. By that time, he had (read extensively)[1] on mathematics, treating it as a fascinating narrative. He became particularly interested in Lagrange's geometry. Lacking good books on algebra, he turned directly to Lagrange and Abel's papers. His personality completely changed—from an active, playful child to a deep, contemplative thinker. His development was constant, and day by day, his teachers grew more puzzled by him, as evidenced in his progress reports.

From 1826 to 1827: "Although this student appears somewhat peculiar, his behavior is impeccable, and his academic performance is excellent."

Later: "Except for the last two weeks, this student came to school because of the fear of punishment. His ambition and creativity have distanced him from his classmates."

From 1827 to 1828: "He is capable and extraordinary."

However, one teacher commented: "I do not agree with the above assessment. I have seen no evidence of his talent, only his eccentricities and indifference!"

Another teacher noted: "He is busy with unrelated matters and is deteriorating."

Later that same year: "He's ruined! He took being mysterious as being creative. His obsession with mathematics has overwhelmed him, leaving him indifferent to schoolwork."

In his final year of school (1828–1829), Galois met a good teacher, Richard, who recognized his genius. Richard noted that Galois was more capable than his classmates. However, when it came to advanced mathematics, he did not complain about Galois not doing elementary exercises. He considered elementary problems to be for ordinary students, while physics and chemistry teachers frequently commented that Galois seemed absent-minded, doing no work at all.

At the age of eighteen, Galois believed he had solved the quintic equation. This was not unusual, as Abel himself, before proving the quintic equation's impossibility, also thought he had solved it. At that time, Galois prepared to take the entrance exam for the École Polytechnique, his greatest aspiration. The École Polytechnique

1 The text in parentheses indicates that the original manuscript is unclear at this point, and the editor has made a judgment call. The same applies below.

was not only a hotbed of mathematical talent but also a cradle of democratic thought. However, he failed the exam, which was a severe blow to him.

The following year (1830, at the age of 19), he published his first paper. He submitted his a long-form manuscript to the Academy of Sciences for the first time, but due to Cauchy's carelessness, the manuscript was lost! This inflicted Galois immense sorrow. He attempted to resubmit it, only to face rejection once again. However, this was not the greatest blow he suffered. A far more devastating tragedy struck when his father, driven to despair by the underhanded attacks of his political adversaries, took his own life in an apartment not far from Galois' residence. As his father's funeral procession passed through his hometown, the citizens who supported him carried the coffin through the streets, while reactionaries jeered from the sidelines. These experiences deeply imprinted upon Galois the sinister and ruthless nature of the reactionaries, leaving an indelible mark on his soul. His aversion to injustice soon evolved into a conviction that only violence could overthrow an unjust system.

Denied entry to the École Polytechnique and mourning his father's death, he found himself lost and directionless. Concerned about his future, he enrolled in the École Normale. Although many students at the school leaned toward liberal ideas, the administration was reactionary, using strict regulations to control students and employing informants to spy on them. Galois endured a double oppression—his political aspirations were suppressed, and his mathematical genius went unrecognized. He was unfamiliar with chalk and blackboard brushes, and his performance in front of the class was consistently poor.

He submitted three papers to a mathematical journal and one lengthy manuscript to the Academy of Sciences. Yet again, misfortune struck—Fourier took the manuscript home but passed away before reading it! Despite extensive searches after his death, the paper was never found. For the young Galois, this was yet another crushing blow. He realized that he was merely another victim of a corrupt system, and his revolutionary convictions grew even stronger.

Then came the gunfire of July 4. One can only imagine how excited this young man must have been! The old regime was crumbling, and the new one was calling. The students of the École Polytechnique rushed into the streets to join the revolutionary ranks, overthrowing the tyrant Charles X. But for Galois, the moment was brutally unfair—his school had locked its gates, barring students from participating.

In the blink of an eye, the glorious three days of the July Revolution had passed. Then, in an opportunistic move, the school's administration released the students. For Galois, however, this lost opportunity was irreparable, and the pain must have been unbearable.

Louis-Philippe had stolen the fruits of the revolution, crushing the hopes of those who truly sought liberty. He maintained a facade of progress while continuing the same oppressive policies as Charles X. Corruption still ran rampant, and the government remained in the hands of the privileged few. During this time, Galois suffered immense pressure—both politically and academically. Everything around him convinced him that revolution was the only way forward.

By the fall of 1830, when he returned to the École Normale, he was a completely different person. Yet the school's principal, instead of understanding the young man's frustrations and offering guidance, constantly reprimanded him—calling him "undisciplined, lazy, ill-mannered, and politically dangerous." The conflict between them escalated until it became irreconcilable. Galois published an open letter in the *Gazette des Écoles*, publicly condemning the principal. This marked the final rupture, and on December 9, 1830, he was expelled from the school. The expulsion was formally ratified by the Royal Council on January 3, 1831.

He rewrote the lengthy manuscript that Fourier had lost and mailed it to *Poisson*. Four months later, Poisson's response arrived—the work was deemed incomplete. This was the final death knell for Galois' academic ambitions. From that moment on, he abandoned all hope of recognition in the scientific community and devoted himself entirely to politics. He had no more illusions of compromise—his resolve to sacrifice himself for the people was unwavering. On one occasion, he openly declared: "If a corpse is needed to stir the people into action, I will offer mine."

On May 9, 1831, at a political banquet, he raised a toast to Louis-Philippe—one hand holding a glass, the other a dagger. Naturally, he was arrested.

His lawyer committed perjury to secure his release. However, on July 14, fearing an uprising, the government once again arrested this dangerous radical Republican. It took the authorities three months to come up with a charge—they accused him of illegally wearing a military uniform. As a result, he was sentenced to six months in prison and incarcerated in the Sainte-Pélagie prison. In 1832, as a deadly epidemic swept through the city, Galois was finally granted bail and released.

But Louis-Philippe's spies were not about to let him escape their grasp. They set a trap. A dubious woman was sent to seduce him, and soon after, a burly man appeared to challenge him, demanding a duel.

On the morning of May 30, 1832, after working through the night, Galois penned farewell letters to his friends and wrote down his final mathematical notes. At dawn, he set off for the duel. Two imposing men awaited him. Galois had no second to stand by his side. The opponent fired—the bullet struck his abdomen, and he collapsed. The duelists left him there. Hours passed before a passing farmer discovered him and took him to a hospital. His only family—his younger brother—stood by his bedside, weeping. Galois told him: "Do not cry. I need all my courage to die at twenty." This was at ten o'clock the next morning.

His funeral was held, attended by two to three thousand Republicans and representatives from the school. A heavy police presence loomed over the procession, fearful of unrest. Yet, the ceremony passed without incident. Those who came to mourn him honored his patriotism, his love for freedom, and his unwavering revolutionary beliefs. But few among them knew that the young political martyr they revered was, in fact, a mathematical genius whose legacy would endure forever.

Friends, as a revolutionary, Galois was forgotten. But as a mathematician, he lives on in eternal memory. He was buried, but his grave has been lost to time. No marker, no headstone remains. He was a genius in mathematics, yet a victim of political injustice. His fate stands as undeniable proof of how corrupt politics can destroy scientists.

Of course, it is not only scientists who suffer destruction from politics. Under Chiang Kai-shek's rule, tens of thousands were slaughtered—among them, perhaps, were many geniuses. The working masses were crushed beneath oppression, their talents stifled before they could ever emerge. Even among the bourgeoisie, how many true talents remained? And if a rare genius did appear, the Chiang regime would either corrupt them or render them mediocre. Students, how fortunate we are to live in a new era! Our political system no longer suffocates us.

We do not need to fight for the future that Galois longed for—we already have it. We should not mourn for him; instead, we should celebrate our own good fortune. Let us dedicate ourselves to our work, so that our full potential may be realized. Let us strive to build a new China!

After the Conclusion of the 1956 Mathematics Competition*

The 1956 Beijing Mathematics Competition has come to an end. The best performances in this competition nearly reached the highest standards we had anticipated. Generally speaking, the vast majority of participants demonstrated the level of outstanding high school students. Achieving such excellent results in the very first mathematics competition is truly exciting. It highlights the extraordinary talent and perseverance of the youth in the new China and signals the emergence of a strong reserve force for the state's scientific endeavors.

Mathematics is the science of quantitative relationships and spatial structures. It is indispensable in all scientific disciplines, as numerical relationships are fundamental to every field. Consequently, mathematics naturally permeates every branch of science. A solid grasp of mathematics provides a firm foundation for entering other scientific domains and enables mathematics to serve a wide range of scientific disciplines effectively. From this perspective, this mathematics competition is not just a source of joy for the mathematics community; it should be recognized as a positive event for the entire scientific community of our nation.

I sincerely congratulate our young participants on their outstanding achievements. However, I hope that the winners will not take pride and complacency from their success but instead become even more humble and diligent, continuing to advance together with their peers. Likewise, I urge those who were not selected as winners not to lose heart or confidence due to a temporary setback. In a long-distance race, the ultimate winner is the one with the greatest endurance.

---

* First published in *Guangming Daily* on May 31, 1956.

At this point, I would like to address some concerns raised by certain junior high school teachers and other friends regarding this competition.

Some have remarked, "The problems were too difficult! Especially those in the second round, which had no equivalent examples in school textbooks." First and foremost, it is important to clarify that the nature of a mathematics competition is different from that of a school examination or even a university entrance exam. Our objective is not merely to assess whether students can apply formulas and theorems but, more importantly, to test their ability to grasp mathematical principles flexibly and apply them to solve problems. In some cases, we even expect students to devise new methods and principles to tackle challenges. Such requirements serve as an effective means to assess and cultivate mathematical talent.

Mathematical talent does not merely consist of memorizing formulas and theorems or solving textbook exercises mechanically. Rather, it involves a deep understanding of mathematical concepts—seeing beyond formulas and theorems to grasp their essence—so that one can apply them flexibly to solve new problems. It is well known that students learning mathematics should develop three key skills: proficiency in computational techniques, a strong intuitive grasp of geometry, and the ability to think creatively. These principles guided the design of our competition. Admittedly, the standards we set may seem high, but it is precisely such challenges that enhance the interest of mathematics enthusiasts and strengthen their mathematical abilities. Moreover, for the true winners of the competition, these problems were not excessively difficult.

On the other hand, the so-called "difficulty" of the problems lies not in their peculiarity or obscurity but in the requirement for a high level of mathematical proficiency. If a teacher misunderstands the nature of the competition and subjects students to a barrage of difficult problems in an attempt to "train" them into winners, we must oppose such an approach, as it is detrimental to young learners. It is evident that merely tackling difficult problems does not yield good results. Even if a student masters one type of difficult problem, they may remain unfamiliar with another. Since difficult problems are endless in variety, how can one possibly prepare for them all? Relying solely on solving bizarre or excessively challenging problems does not cultivate true mathematical talent.

Only by mastering fundamental principles can one approach problems with confidence and adaptability, allowing for true creativity and flexible application. The application of principles is not limited to solving a fixed set of problems—it extends to a broad range of challenges in various forms. Thus, attempting to win by simply practicing numerous difficult problems is neither an effective nor advisable strategy. I encourage both teachers and students to focus on teaching and learning fundamental concepts rather than getting lost in exhaustive and often fruitless problem-solving exercises. Of course, moderate exposure to challenging problems remains beneficial as a means of practice.

There is also the voice that the competition lacked calculation-based problems. Indeed, this is a shortcoming. Testing computational skills matters a great deal, as interesting calculation problems cannot simply be solved by mechanically applying formulas or fixed methods—they require intelligence and skill. However, finding suitable calculation problems that demand skill while also being solvable within a short timeframe is quite difficult.

Let me clarify once again: The absence of calculation problems in the competition does not mean they are unimportant. On the contrary, in all practical applications, one must arrive at concrete numerical results rather than merely substituting values into a formula. Calculation is a crucial link between mathematical theory and other sciences—an indispensable element. Therefore, students should develop strong computational skills and not neglect them just because they may seem tedious or monotonous. Otherwise, they may face significant difficulties in their future scientific pursuits.

The top-ranking winners of the mathematics competition have received many letters—some offering congratulations, while others express dissatisfaction and issue "challenges" such as: "Let's see who is truly better!" I believe that such challenges, born out of dissatisfaction, are neither beneficial nor necessary. It is natural for young people to be determined and unwilling to accept defeat, and this spirit is commendable. However, determination should not come at the expense of humility and a commitment to truth.

Those who did not win should adopt the right attitude—first, by analyzing the reasons for their defeat and striving to improve, and second, by congratulating the winners, respecting their achievements, and learning from their success. Every individual accomplishment should be regarded as a collective honor, a reflection of the brilliance of one's peers. Jealousy is unproductive. One should not fear that their talents were not fully displayed in this competition; rather, the real concern should be losing confidence or failing to put in the necessary effort. Who can say that among the students who did not qualify this time, there won't be future talents who surpass today's winners?

At this point, I would like to convey a request from young students to the scientists.

During this competition, we organized a series of lectures, which were met with great enthusiasm. Many students eagerly hope these lectures will continue. At one public lecture, some students approached me and expressed their strong desire for scientists from various fields to hold more talks to enrich their scientific knowledge. They repeatedly urged me to convey this message to our esteemed scientists. Their sincerity and eagerness are so moving that one cannot help but be determined to fulfill their wishes, no matter what. Supporting and guiding young minds in this way aligns perfectly with both our responsibilities and aspirations as scientists. We have a responsibility to nurture the growth of the next generation, to continually offer them intellectual sustenance through diverse channels, to stay closely connected with them, and to provide guidance in all aspects of their development. By doing so, we can ensure that our successors are better prepared for what's required of them. I deeply resonate with the eager hearts of the youth, and therefore, I humbly urge our distinguished scientists: Meet with them often, offer them concrete guidance, and help fulfill their passionate thirst for knowledge!

This year, only four cities—Beijing, Tianjin, Shanghai, and Wuhan—held mathematics competitions. Many students from other cities have written in, expressing their hope that this initiative can be expanded to their regions. Their enthusiasm for scientific endeavors is truly encouraging. However, due to our lack of experience and limited resources, we were unable to meet this demand this year. Next year, we are determined to expand the competition appropriately, and we are confident that we will organize it even better than this year. To achieve this goal, we will need the support of all sectors.

Intelligence Lies in Learning, Genius Comes from Accumulation (1956)*

Recently, the Party issued a solemn call for a major push into science, asking us to reach the advanced global level in key scientific fields within twelve years. This call has greatly encouraged young scientific workers, and many young people have even set up several years' worth of study plans. This is a very promising development. Here, I would like to offer a few suggestions for everyone's consideration.

■ Intelligence Lies in Learning, Genius Comes from Accumulation

We must understand the long-term and arduous nature of attacking the fortress of science. Like an army in battle, to conquer a strongly fortified position, it is not only about charging in forcefully, but also about understanding strategy and tactics. Advancing into science requires not only bold imagination and a constant dissatisfaction with current achievements but also a steady, practical approach to small tasks, along with long-term hard work.

I have heard that many university graduates are setting plans to earn an associate doctorate in a few years. But I advise everyone not to believe that once you get an associate doctorate, all will be well, and don't think that once you push for a doctoral degree, you can stop researching. No, scientific research is a lifelong commitment. Our task is to build a happy communist society, explore the mysteries

* First published in issue 7 of *China Youth* in 1956.

of the universe, and make the forces of nature serve humanity. This is an endless mission. If one relies solely on a few months or two to three years of intense effort and then stops, it is hard to expect any significant achievements. Even if one achieves some results, it will not reach the peak of science, and any success will be limited and very small.

Before the liberation, we saw many scientists whose careers were marked by a path that went from university graduation to studying abroad, then from studying abroad to earning a doctorate, and then from a doctorate to becoming a professor. They might have had ambitious aspirations to reach the heights of science when they were at university and learned some useful knowledge while studying abroad. There might have been some valuable or creative work in their doctoral theses, but once they earned their doctorate and became professors, they stopped. They abandoned scientific research, and decades later, they could not produce a single decent paper. This is truly regrettable. Of course, this was largely due to the evil environment of the old society. Today, our environment is different. The socialist system of the People's Republic of China has opened up vast opportunities for science. Today, we are able to wholeheartedly dedicate ourselves to our roles, immersing ourselves in scientific endeavors, and passionately pursuing research and innovation. Our scientific work has become an inseparable part of the entire socialist cause. Therefore, we should no longer see science as a "stepping stone" but set the highest standards and goals for ourselves, and stick to them. The more we create and the more profound our contributions to the people, the better.

Some comrades lack perseverance and determination because they have encountered setbacks and taken wrong turns in their work, which causes them to doubt their ability to engage in scientific research. In reality, I can assure you that many famous scientists and writers have experienced numerous failures and walked many detours before finding success. People often marvel at a writer who crafts a brilliant novel or a scientist who publishes groundbreaking papers, yearning to effortlessly produce such masterpieces themselves, as if they could wake up one day and be celebrated by all. In truth, successful papers and works represent but a small fraction of an author's entire body of creation and research. Often, the number of these successful pieces pales in comparison to the countless drafts and failures, sometimes accounting for less than one-tenth of the whole. What we see is only their successful works, while the failures are never publicly shared. It is essential to recognize that in a scientist's long and challenging journey to conquer the fortresses of knowledge, the lessons drawn from failure often surpass those gleaned from success. While failure is never a joyous experience, it should never be a reason for discouragement. It is common for researchers to stray from their course, expending energy on paths that lead nowhere. Yet, this is not the end. One can always recalibrate and pursue a more accurate direction. The key lies in embracing failure as a valuable teacher—learning from it, reflecting on the insights it offers, and using these lessons to press forward.

Based on my personal experience, so-called genius is the result of persistent effort. Some comrades may think I have some innate talent for mathematics, but in fact, there is no trace of such talent in me. When I was in elementary school, I didn't receive my graduation certificate due to poor grades and could only get a completion certificate. In the first year of junior high school, I barely passed my math exams after a retake. However, strangely enough, starting from the second year of junior high school, a fundamental change occurred. This was because I realized that since I didn't have natural talent, I needed to dedicate more time to studying. While others studied for one hour, I studied for two, and gradually my math performance improved. To this day, I still follow this principle: While others might take three hours to study something, I spend three and a half hours. After long periods of accumulated hard work, I've been able to achieve better results. Once the basic skills become second nature, I can often read an article that others might take ten days or two weeks to fully comprehend in just one hour. Therefore, the extra effort invested in one period will yield unexpected results in another. Yes, intelligence lies in learning, and genius comes from accumulation.

■ Grounded and Accelerated

Since scientific work is a long-term and arduous endeavor, it not only requires perseverance and determination but also demands attention to scientific methods and procedures. We must proceed step by step, gradually and steadily, just like how we cannot expect to reach the beautiful future of socialism in one morning; a transitional period is necessary. Advancing in science is like climbing a ladder—you must step up one step at a time, both steadily and quickly. If you attempt to leap over several steps at once, it will inevitably result in a fall and cause great setbacks. Am I being conservative in my thinking? Does this contradict the principle of "more, faster, better, and more efficient"? I believe that gradual progress is not in conflict with speeding up. Because of gradual progression and solidifying the foundation, we can ensure smooth progress in the long run. In the past, some junior high school students, considering themselves geniuses, skipped several grades, and some even left some junior high school prematurely to jump straight into high school. This is a very dangerous thing. Although they may keep up temporarily, their lack of a solid foundation will cause great difficulty in further research. Some young people not only fear difficulty but also belittle the easy things. For instance, they may not take some junior high school arithmetic seriously before jumping to algebra, thinking that arithmetic is too simple to spend time on. However, when they face algebra, they realize many things they cannot understand, which leads to considerable difficulties. In fact, the difficulties we often face are simply due to our underestimation of seemingly simple things.

I myself have experienced such pain. When reading a thick book, the first couple of chapters always seem easy, I study them quickly and move on carelessly. But by the time I reach the third or fourth chapter, I feel the material becoming more difficult. By the fifth or sixth chapter, I find it hard to continue. If I don't want to give up halfway, I have to go back and review the earlier chapters more carefully. Of course, when I say to proceed gradually and solidify the foundation, I do not mean to endlessly circle around the same place, re-reading the same books repeatedly. For example, in the past, some people researching mathematics collected several textbooks on calculus of the same level, read each one from beginning to end, and even repeated the exercises in them. This is a bookish approach, with no real practical meaning, and would naturally violate the principle of "speed." My personal view is that when laying the foundation of knowledge, it's sufficient to choose one good book under the guidance of a teacher, read it carefully, and finish it (and subsequent readings of similar books should focus on absorbing new material rather than re-reading the entire book). Then, move on to more advanced books. Gradual progress should never mean staying at the same level, but should involve stepping forward, and doing so quickly, one step at a time.

Regarding the issue of strengthening foundational knowledge, there are currently two viewpoints in universities: One is to work, research, and simultaneously strengthen the foundation; the other is to solidify the foundation first before researching. Both methods can achieve gradual progress. However, which method is better must be decided based on specific conditions and environments; it cannot be applied mechanically. I believe that with a good mentor providing guidance, it's acceptable to work on both research and strengthening the foundation simultaneously, as this avoids unnecessary detours and enables rapid progress. Without a mentor, however, it's essential to first solidify the foundation. Without a solid foundation and no guidance, it's easy to encounter difficulties when conducting research and may lead to abandoning the research halfway or achieving less-than-expected results. But even without a mentor, the time spent on strengthening the foundation will not be long. I've heard that some university graduates, after teaching for two or three years, still plan to spend ten years strengthening their foundation and striving for an associate doctorate. This is completely unnecessary. In my personal opinion, two or three years is enough for a student who has completed their third year of university and enters the workforce to strengthen their foundation. Of course, this refers to two or three years of diligent effort, not inconsistent, sporadic effort.

■ Independent Thinking and the Pursuit of Rigorous Training

One of the most critical factors in achieving excellence in scientific research is the ability to think independently. Scientific undertaking is, at its core, a creative endeavor. Our mission as researchers is to uncover what has yet to be discovered, to push beyond the frontiers of knowledge. If everything were already known and neatly explained, there would be no need for research at all. That is why simply adopting others' experiences is insufficient. The natural world is in constant flux, and science advances at an ever-accelerating pace. Relying solely on others' experiences can leave us incompatible with the work. Even our own past experiences may not always serve as reliable guides when confronting new challenges. Instead, we must continuously weed through the old to bring forth the new, and boldly create.

I have always believed that many young scholars in our country struggle with this aspect of independent thinking. During a visit to East Germany, I spoke with Chinese students studying there. They told me that because they had not been trained to think independently in Chinese universities, they faced significant difficulties in their studies. While they performed just as well as their German classmates in lectures and reading assignments, they found themselves at a loss during seminars in which they were expected to engage critically with ideas. Some even had trouble finding relevant research materials, and when they did find them, their presentations often consisted of little more than repeating what others had already said, without adding their own insights or making original contributions.

Indeed, I have met many university students who never even consider challenging what is written in their textbooks. As a result, they become little more than passive conduits of knowledge. Some Chinese universities have historically promoted a "spoon-feeding" approach to education. If a student does not understand something the first time, they ask the teacher; if they still don't understand, they simply keep asking—until they grasp the material. While this may seem like an efficient learning method, the unfortunate reality is that no discipline can be mastered through passive instruction alone. If a teacher were to provide all the answers, there would be no need for the student to conduct research in the first place. A mentor's role is to offer guidance and point students in the right direction, preventing them from wandering aimlessly. But the specific challenges and obstacles along the way must be confronted and overcome by the students themselves.

This is especially true given the many scientific gaps that still exist in China. In numerous fields, no one has yet undertaken the necessary research. If you rely solely on others to guide you, who will you turn to for answers? The only viable path forward is to build upon the knowledge you already have, to think critically, and to innovate with careful deliberation.

However, let me be clear: Advocating for independent thinking does not mean disregarding the wisdom of those who came before us and cockily doing it on our own. Independent thinking is not the same as blindly stumbling through research without seeking guidance. If someone lacks a solid foundation in scientific knowledge yet insists on "thinking independently," shutting themselves away to rely solely on personal contemplation, they may come up with ideas with their talent—but the chances are high that these ideas will have been discovered decades, if not centuries, earlier. Their efforts would be wasted, and they would certainly fall behind the global pace of scientific progress.

Therefore, it is essential to learn from the experience of predecessors and absorb the scientific achievements already made by the world. To accomplish this, young scholars should actively seek guidance from experienced mentors and undergo rigorous training.

■ Practice Makes Perfect

Finally, I would like to briefly discuss two methodological issues. In my opinion, the most fundamental principle in any method is that practice makes perfect. Mastery is essential in any field—only through deep familiarity can one innovate and make discoveries. However, the mastery I refer to does not mean rote memorization of formulas, laws, or pre-existing conclusions. Simply memorizing something does not necessarily mean true understanding, and true understanding does not always require memorization.

If someone reads the same book ten or twenty times without truly comprehending or applying its concepts, that is not mastery—it is mere recitation. Mastery means grasping the core links of your field, understanding how past scholars thought and discovered these ideas. For instance, if an experiment requires five steps, one must understand why all five are necessary—whether one can be omitted, and how previous researchers arrived at these steps. This kind of thinking is crucial because the goal of scientific research is to invent or discover something new. If someone presents you with an invention and you cannot even understand the process by which it was created, how can you hope to further innovate and develop something new?

Take ceramics as an example: If we do not understand how traditional porcelain is made, how could we hope to invent a new kind? In capitalist countries, there is a tendency to mystify scientific discoveries, as seen in the widely told (and misleading) story of Newton and the apple. It is often said that Newton discovered the law of universal gravitation simply because he saw an apple fall and had a sudden inspiration. This is nonsense. Apples had been falling to the ground for as long as humans have existed—so why did so many people see it happen without making the same discovery? The truth is, Newton did not just observe a falling apple; he was already deeply engaged with Kepler's laws of planetary motion and Galileo's laws of falling bodies. After years of contemplation, when he encountered a natural phenomenon, he was able to immediately perceive its underlying principles.

Therefore, one must have a thorough understanding and mastery of the process by which key theorems are derived in order to make meaningful progress in science. To clarify, the "key" discussed here does not refer to every possible problem's key point but rather to the primary critical aspects within the specific field of research one is engaged in.

Then, there is the issue with materials. Since scientific research involves building on the experience of predecessors, it is crucial to have access to comprehensive materials. If you are researching an underdeveloped area—one that has little to no prior research in China—finding references will be far more challenging.

Rather than discussing abstract principles, I prefer to offer a concrete approach, though I acknowledge that the more specific the advice, the greater the chance of it being flawed. Please take my suggestions with discretion and adapt them as needed.

If you have a mentor, they can guide you by pointing out previous researchers in the field and recommending key reference materials. While they may not have all the answers, their guidance can provide a valuable starting point, which is easier. However, if you are tasked with establishing an entirely new research area on your own, what should you do?

First, identify the most authoritative figures or schools of thought in your field worldwide. Then, study their most recent publications. Initially, you may struggle to understand them. This could be due to two main reasons: First, they reference textbooks you have never studied. This is a valuable insight—it reveals gaps in your foundational knowledge that you need to address. Second, they cite numerous other researchers' works. While you do not need to read all of them, you can use these citations as a roadmap, expanding your knowledge step by step from their references.

By following this method, you can build a solid foundation efficiently—some may take one or two years, while others might need three to five years to grasp the full scope of their field.

On Junior High School Mathematics Textbooks*

The significance of mathematics is universally recognized. Wherever quantitative relationships exist, mathematics plays an essential role. Any discipline that involves quantities and their interrelations, and seeks to depict them with precision, inevitably encounters mathematical problems. Once a problem is framed mathematically, the methods and conclusions of mathematics become valuable assets to that field of study. Thus, mathematics serves as an excellent tool for exploring any branch of natural science. Mastering it facilitates entry into other domains of natural science.

On the other hand, mathematics cannot replace other natural sciences. Its value lies in its applications. A principle or formula, when applied to different contexts, takes on different meanings and functions. If mathematics is detached from its applications and studied in isolation, it loses its significance and can no longer be assessed for its importance. Mathematics itself is neither inherently important nor unimportant—it gains its importance from its wide-ranging applications, its omnipresence, and its precision and reliability.

If students learn arithmetic but cannot convert between different units of measurement, or if they study geometry but cannot calculate areas, this is deeply concerning. Such a disconnection—where knowledge cannot be immediately applied to real-world situations—should not be allowed to persist. For students, forming this bad habit would be a tremendous loss for the socialist construction. To put it more seriously, it fosters an environment where theory is detached from practice.

I am not thoroughly familiar with the junior high school textbooks, so the following opinions may be imperfect.

It appears that we allocate twelve years in primary and secondary education to cover material that is taught in ten years in Soviet schools. This approach does not

---

* First published in issue 7 of *Junior High School Teachers* in 1958.

align with the principle of "more, faster, better, and economical." I am uncertain why primary and secondary education in our system takes twelve years. Some suggest that our written language is complex and thus requires more time to learn. Even if that is the case, why must we take twelve years to teach the same mathematics curriculum that Soviet schools complete in ten? Mathematics is equally difficult for both Chinese and Soviet students, so why should we spend an extra two years?

Thus, my tentative suggestion is that if we maintain the twelve-year education system, we should ensure that the mathematics curriculum genuinely reflects the extended duration. In other words, beyond covering the ten-year Soviet curriculum, we should add two additional years of mathematical content.

Regarding which subjects should be added, my first thought is analytic geometry. Why do I make this recommendation? Because this subject forms an organic synthesis of elementary algebra and geometry. It employs algebraic methods to study geometric properties and uses geometric figures to illustrate the significance of algebraic expressions. Furthermore, it serves as a powerful tool for representing objective phenomena graphically and analyzing the rise and fall of real-world systems. This subject is both a culmination of junior high school mathematics and the first step into advanced mathematics.

Moreover, the absence of analytic geometry in secondary education creates challenges for university-level instruction. For example, the study of calculus must be based on analytic geometry, and university physics, in turn, requires the use of calculus. Since university physics is typically a first-year course in higher education, the curriculum cannot be arranged optimally, creating significant difficulties for both science and engineering students. If analytic geometry could be covered at the secondary school level, it would greatly enhance the effectiveness of mathematics instruction in higher education.

As for whether an introduction to calculus could also be added to the secondary school curriculum, this would depend on the overall structure of the mathematics syllabus and the availability of qualified teachers. However, one possible approach would be to introduce fundamental concepts of limits and the essence of differentiation in an enlightening way while teaching analytic geometry. This would provide students with a foundation for further study of calculus.

Another topic worth considering for inclusion is an introduction to probability (sometimes referred to as "chance" or "likelihood"). There are two primary ways to describe the objective world mathematically: One focuses on deterministic laws, typically using equations and equalities; the other describes large-scale phenomena, with probability as its key tool. Gaining an early understanding of probability and mastering its basic computational principles is highly important. This subject is not particularly difficult to learn and provides significant insights into non-deterministic phenomena. It has practical applications in fields such as genetics and selective breeding, as well as quality control in manufacturing.

Additionally, another topic that could be introduced is Qin Jiushao's method in algebra (known in the West as Horner's method). This algorithm is not only a brilliant achievement in the history of ancient Chinese mathematics but also remains a fundamental method for solving algebraic equations today. I have encountered many students who have studied advanced mathematics yet cannot compute the fifth root of 2. This exposes serious gaps in their mathematical knowledge and is a disservice to our ancestors who discovered such methods over a thousand years ago.

Furthermore, some propose adding determinants and elimination methods to algebra, as well as incorporating more applied measurement calculations in plane geometry, solid geometry, and trigonometry. I find these suggestions highly valuable. Developing students' spatial intuition during solid geometry instruction is also particularly important.

Moreover, I would like to emphasize that junior high school students must develop the habit of critical thinking and accurate computation. From what I have observed, this aspect does not seem to receive enough attention. Students must understand that without computation, mathematical knowledge cannot be effectively applied to real-world situations. To ensure that their mathematical learning is not wasted and that mathematics education remains closely tied to the country's socialist development, we must instill in students the habit of critical thinking and accurate computation from an early stage. Of course, by computation, I do not mean only written calculations, but also proficiency in mental arithmetic and abacus calculation. The abacus, in particular, is a highly practical computational tool unique to China. Mastering its use can bring great convenience in real-world applications. Similarly, critical thinking should not be limited to solving difficult mathematical problems but should extend to formulating real-world problems in mathematical terms.

Regarding the repetition of certain topics in mathematics textbooks across different education levels, I have some preliminary thoughts. I agree that simple repetition lacks value, but higher-level reinforcement is both inevitable and necessary. For instance, after learning algebra in junior high school, revisiting algebra from the beginning in high school should not be considered a waste of time. At that stage, reviewing earlier concepts requires only 1/10 of the original study time yet helps students consolidate and refine their knowledge. This process should make textbooks feel "thinner" rather than "thicker," leading to a deeper understanding rather than an accumulation of disorganized information.

On the other hand, certain sections of the junior high and high school curriculum could be streamlined. For instance, excessive time spent on complex geometric problems might be unnecessary, and arithmetic and algebra could be more closely integrated. Specific details on these aspects need further discussion, but I will not elaborate on them here.

The Great Utility of Mathematics*

■ Number and Quantity

The concept of numbers originates from counting, as seen in one, two, three, four, five … or one unit, two units, three units, and so on. Similarly, the concept of quantity originates from the act of measuring, which involves taking a standard unit and measuring one unit at a time. Though there are various units of measurement in the world—such as meters, kilograms, liters, seconds, volts, ohms, and calories—all must be expressed through numbers to accurately represent real-world conditions. Thus, numbers are the common property of all forms of measurement, enabling the comparison of quantities and the description of their variations.

Since quantity permeates all scientific fields, the application of mathematics extends across every scientific domain. Whenever there is a need to study quantities, their relationships, their changes, or the variations in these relationships, mathematics becomes indispensable. Moreover, such changes themselves can undergo further changes, which are typically expressed quantitatively. For example, speed describes the motion of an object, while acceleration captures the change in speed. The relationships between quantities are diverse, and even different types of relationships

* This article was first published in *People's Daily* on May 28, 1959. It was later reprinted under the title "The Applications and Development of Mathematics" in *An Introduction to Modern Science and Technology* (Science Press, 1978). At the time of reprinting, the author recognized the significant advancements that had taken place over the years and felt that the content needed revision and expansion. However, due to time constraints, only a refined version of the original manuscript—based on the author's dictated notes—was published. The author earnestly hoped to receive insightful feedback from various perspectives, allowing for further refinement and enrichment of the article in the future.

may be interrelated. Among these relationships, some are fundamental, while others can be derived from them. Therefore, mathematics does not merely study quantities but also explores changes in change, relationships between relationships, and commonalities among commonalities—advancing in cycles, continuously refining itself, and progressing indefinitely.

Mathematics serves as a powerful tool and assistant to all sciences. At times, its development is driven by other scientific disciplines, while at other times, it takes the lead in advancing knowledge, later finding its applications. Without mathematics, no scientific discipline can precisely depict the changing states of objective reality, nor can it derive unknown data from known data. This would limit the potential for scientific prediction or at least reduce its accuracy.

As Engels stated, "Pure mathematics deals with the spatial forms of the real world and the quantitative relations therein." Mathematics is an abstraction derived from physical models, encompassing both numerical relationships and spatial structures. The discussion thus far has focused on quantitative relationships—now, let us explore spatial forms in relation to the vastness of the universe.

■ The Vastness of the Universe

The vastness and form of the universe can only be accurately described through mathematics. The ancient belief that "the heavens are round and the earth is square" was an early attempt to depict the cosmos using geometric models. This notion—likening the sky to a domed canopy and the land to a chessboard—was eventually disproven by navigators through empirical evidence. However, theoretical skepticism toward this model emerged much earlier in China and was equally compelling. One argument stated: "At the dawn of chaos, heaven and earth were established. The light and pure elements floated upward to form the sky, while the heavy and turbid elements condensed below to form the earth." But this raised further questions: What lies beyond the realm of the light and pure? What exists beyond the celestial dome? What lies beyond the so-called "thirty-three heavens"? To answer such questions, one must rely on mathematical studies of spatial structures.

The concept of four-dimensional space may sound mysterious, yet it has existed for centuries. The very term universe reflects this idea: "The past and the future define time, while the four cardinal directions and vertical dimension define space," as stated in *Huainanzi: Lessons of the Common Folk.* This means that space represents three-dimensional expansion—length, width, and height—while time represents a one-dimensional temporal flow. During Newton's time, this understanding of the universe prevailed: an infinite three-dimensional space within which celestial bodies moved. Newton's great achievement was uncovering the laws governing these motions—a prime example of how physical models drive the development of mathematical methods, which in turn help explain physical phenomena.

Since planetary motion does not follow a uniform acceleration, describing such motion necessitated a means to analyze continuously changing velocity. This led to the development of calculus, an invaluable tool for characterizing acceleration. Newton, wearing two hats as both a physicist and a mathematician, not only formulated calculus but also discovered the law of universal gravitation. With these tools, the motion of all celestial bodies in the universe was initially unified and explained—one could calculate why planets orbit the sun in ellipses, their speeds at any given moment, and their precise locations at any point in time.

Some attribute the rapid advancement of Western civilization to two key factors: Euclidean geometry's logical reasoning system and the scientific method of systematic experimentation. Newton's work exemplified rigorous logical deduction—deriving the entire framework of mechanics from a few fundamental laws. His theories explained everything from celestial orbits to the construction of bridges, buildings, and levers. Yet, the theories that people establish in understanding the natural world are never final or flawless; Newtonian mechanics, for all its success, left some questions unanswered. It explained planetary orbits but not planetary rotation. For instance, earth completes a full rotation in 24 hours, resulting in day and night, whereas Mercury's rotation matches its orbital period, leaving one hemisphere in perpetual daylight and the other in eternal darkness. A particularly puzzling issue was Mercury's anomalous precession—its orbit shifts by 42″ per century, a phenomenon Newtonian mechanics failed to explain.

Einstein no longer viewed space and time separately; rather, time was also in motion. At every moment, matter occupies a specific position within three-dimensional space. Building upon Maxwell-Lorentz's postulate of constant light speed and Newton's principle of relativity, he formulated the Special Theory of Relativity. The Lorentz transformation in special relativity intertwines space and time, yet it does not erase their distinct characteristics. For example, if one walks three miles east and then three miles west, they return to their starting point—but time, unlike space, does not rewind; it flows irreversibly forward.

It is worth noting that some have calculated that special relativity not only fails to explain Mercury's perihelion advance but even predicts a "retrograde motion." This is a misunderstanding. Our calculations show an advance of 28″, which accounts for two-thirds of the observed value. Moreover, with a deep analysis, the explanation becomes more intuitive—one can derive all the results of Einstein's special relativity with fewer assumptions and even without using calculus.

Einstein then extended his theory by linking space, time, and matter, formulating the general theory of relativity. Using this framework, he successfully calculated Mercury's precession as 43″, providing strong evidence in support of general relativity. Since the available evidence is still limited, there are various perspectives on general relativity. However, its development relied on mathematics taking a step ahead, as seen in the prior existence of Riemannian geometry. On the other hand, general relativity has also presented mathematics with numerous unresolved problems. How much further our understanding of the universe will advance remains unknown, but one thing is certain—every step forward will inevitably depend on mathematics as a crucial tool.

■ The Minuscule World of Particles

Buddhist scriptures speak of a "Golden Millet World," suggesting that even a single grain of millet could be regarded as an entire world. While this is a product of Buddhist imagination, modern science has revealed that atoms—far smaller than a grain of millet—possess complexities comparable to a miniature solar system.

Even the study of atomic nuclei, which are vastly smaller, cannot be conducted without mathematics. Describing the motion of fundamental particles within the nucleus is even more dependent on mathematical tools. Can we apply the same mathematical methods used to study the macroscopic world to the internal structure of nucleons? The situation is different. Here, both Newtonian mechanics and Einstein's relativity encounter significant challenges. In response, modern physics has turned to new mathematical frameworks, such as operator theory, group representation theory, and generalized function theory. These are all products of recent mathematical advancements. Yet, even with these tools, a complete understanding of atomic structure remains elusive.

Whether examining molecular theory, atomic theory, or the latest studies on nucleon structures and particle interactions, the evolution of physical science has always been inseparable from mathematics. This has been true in the past and will undoubtedly remain true in the future. One certainty is that mathematics will continue to be essential.

One of the major unresolved questions in physics today is whether a unified framework can be developed to reconcile both the macroscopic and microscopic worlds and integrate the four fundamental forces of nature into a single theory. Regardless of how physicists ultimately solve this problem, the mathematical methods required must be unified. A mathematical system must be developed that can simultaneously explain both the vast cosmos and the subatomic realm. Just as mathematics has always been the indispensable tool of physics since its inception, it will continue to play a crucial role in these grand scientific endeavors. Moreover, just as Newton simultaneously developed celestial mechanics and invented calculus, the pursuit of solving such a grand problem may lead to the creation of entirely new branches of mathematics.

■ The Speed of Rockets

In today's world, the phrase "traveling a thousand miles in a day" is better suited to describe slowness rather than speed. The velocities that humans can achieve far exceed this. While modern aircraft can traverse thousands of miles in a day, their speed still pales in comparison to the vast scale of cosmic velocities. The ancient fantasy of traveling from New York in the morning to Los Angeles by evening is no longer a marvel in our era.

Reflecting on the early days of interstellar travel—when poetic dreams first transitioned into scientific reality—mathematics was indispensable from the very beginning. As early as Newton's time, the first cosmic velocity, approximately eight kilometers per second, had already been calculated, setting a clear goal for scientists and engineers. Achieving this speed would make it possible to launch satellites into earth's orbit. In 1970, China successfully launched its first artificial satellite, a feat in which mathematicians played a crucial role (though many of them were recognized not as pure mathematicians but as experts utilizing mathematical tools).

To place an artificial planet in orbit around the sun, a velocity of 11.2 kilometers per second—known as the second cosmic velocity—is required. To escape the solar system and venture into interstellar space, a speed of 16.7 kilometers per second—known as the third cosmic velocity—is needed. These ambitious goals are progressively being pursued and realized.

On a related note, suppose our spacecraft reaches another planet inhabited by intelligent life forms similar to humans. How would we establish communication with them? Bringing a painting would be futile, as they might not comprehend our artistic perspective. Playing an audio recording would be equally ineffective, as language barriers would prevent understanding. The best approach, I believe, would be to present two fundamental mathematical representations: one showcasing numerical concepts and another illustrating numerical relationships, such as the Pythagorean theorem (fig. 1 and 2).

To further demonstrate our capability for geometric reasoning, we could also provide the "Green Out, Vermilion In Diagram" (fig. 3), an achievement from ancient Chinese mathematics.

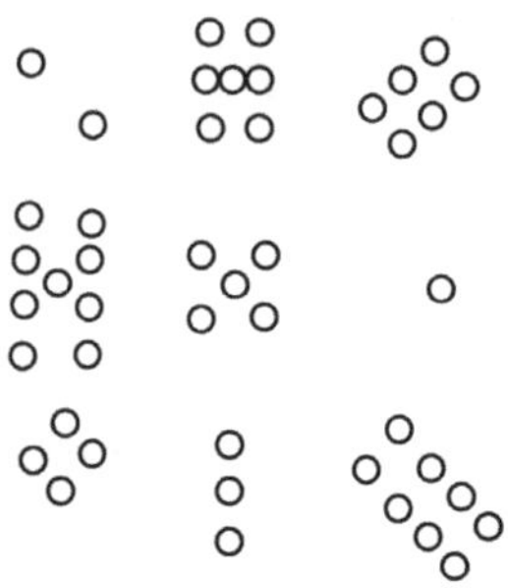

Figure 1

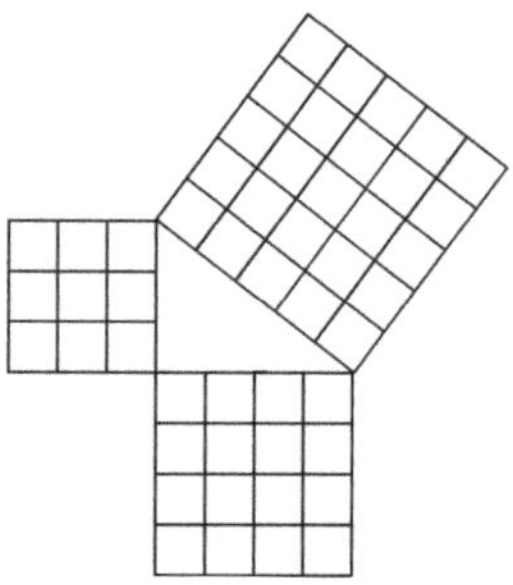

Figure 2

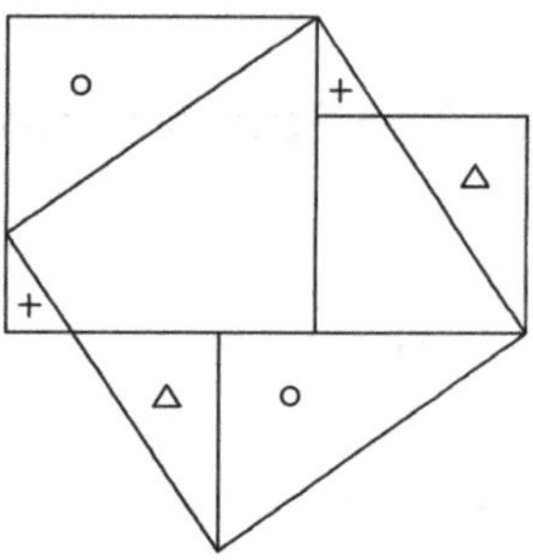

Figure 3

The Ingenuity of Chemical Engineering

The vast array of new products created by the chemical industry has greatly enriched human life, truly embodying the phrase "surpassing the craftsmanship of nature itself." In the manufacturing process, chemical reactions—both synthesis and decomposition—are described by chemical equations. However, since these reactions are constantly evolving, the great revolutionary thinker Engels pointed out: "All chemical equations that represent the molecular composition of substances, in their formal structure, are differential equations. However, these equations have effectively been integrated due to the atomic weights they represent. What chemistry calculates are precisely the differentials of known quantitative relationships."

To illustrate this concept, consider the extraction of sulfur from a substance using benzene. Benzene has a saturation limit for absorbing sulfur. As benzene absorbs more sulfur, its ability to absorb additional sulfur decreases. Similarly, as the remaining sulfur diminishes, it becomes increasingly difficult for benzene to extract more. This process is continuously changing, with the absorption rate gradually slowing down. The experiment itself represents an integration process, and its mathematical representation is a differential equation and its solution.

For simple processes, experimental methods may be sufficient to obtain results. However, for complex and challenging processes, mathematical techniques are often necessary. In particular, optimizing industrial processes relies heavily on mathematical tools. With the advancement of quantum chemistry, chemical research has progressed to the realm of quantum mechanics, where mathematical methods—such as differential equations, matrices, and graph theory—are indispensable.

The application of mathematics has also greatly simplified theoretical problems in chemistry. For instance, calculating energy levels in conjugated molecules becomes increasingly difficult as the size of the molecule grows. However, using molecular orbital graph theory simplifies these calculations, providing a more intuitive and feasible approach. The underlying principle is that the more zeros present in the determinant of a system, the easier it becomes to simplify calculations using graph theory.

■ The Changing Earth

The earth we live on is in a constant state of flux. From the upper layers of the atmosphere to the vast oceans, down through earth's crust and into its core, intense movements are always occurring. The study of these dynamic processes relies heavily on mathematics.

Atmospheric circulation, wind patterns, and precipitation require continuous research and forecasting. Accurate weather predictions allow farmers to plan agricultural activities and enable safe air travel. Forecasting extreme weather events, such as hurricanes, helps naval forces, fishermen, and coastal communities prepare in advance, reducing potential damage. All these predictions are fundamentally dependent on mathematics.

The relationship between wind and water has been observed for centuries. The old saying, "A sudden wind ripples the surface of a spring pond," captures the connection between air currents and water movement. Another common phrase, "No wind, no waves," hints at the link between wind and ocean waves. However, a precise scientific understanding of this relationship has only emerged in recent times, utilizing mathematical tools such as discontinuous solutions in partial differential equations.

Each year, millions of earthquakes occur worldwide. Most are too small to be felt, and even larger ones may not cause disasters if they strike uninhabited regions. However, a few major earthquakes each year affect densely populated areas, leading to devastating consequences. The study of crustal movement, earthquake prediction, and even future efforts to control seismic activity all depend on mathematical analysis.

■ The Mysteries of Biology

Biology is filled with mathematical problems. Consider the structure of a honeycomb: Why do honeybees construct their hives in this particular shape (fig. 4)? The front and back faces of a honeycomb cell appear as perfect hexagons, but the structure itself is not simply a hexagonal prism. Instead, the bottom of each cell is formed by three rhombuses. Figure 5 provides a clearer and more specific three-dimensional illustration of this structure.

To better understand this, consider a hexagonal prism, like an unsharpened pencil. Looking at its cross section, one end forms a regular hexagon, labeled *ABCDEF* (fig. 6). If you cut off one corner along *AC* and move the triangular piece *ABC* to position *AOC*, then make similar cuts along *AE* and *CE*, you form the shape shown in figure 7. A honeycomb is essentially two layers of such structures joined at their bases.

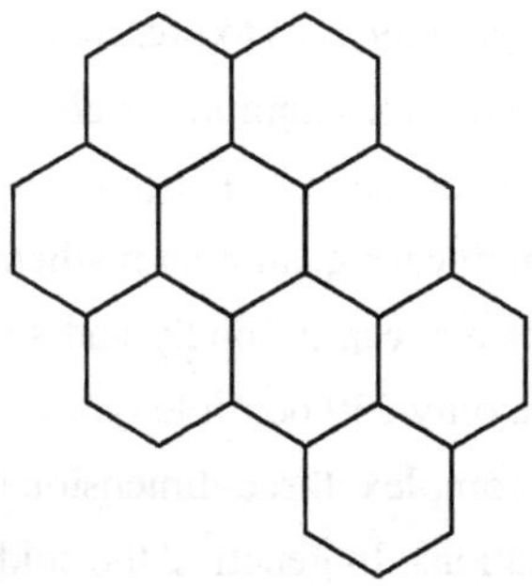

Figure 4

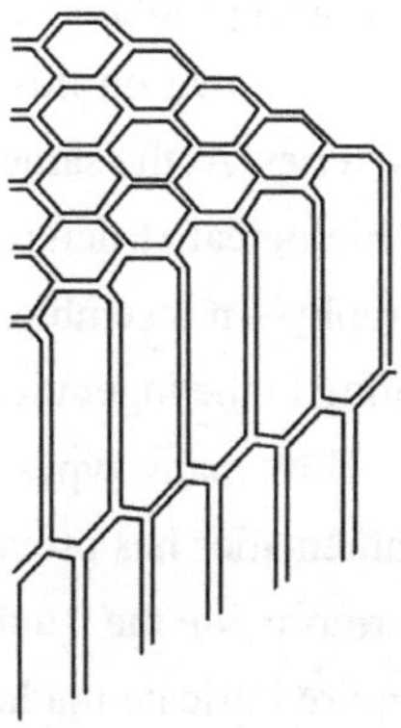

Figure 5

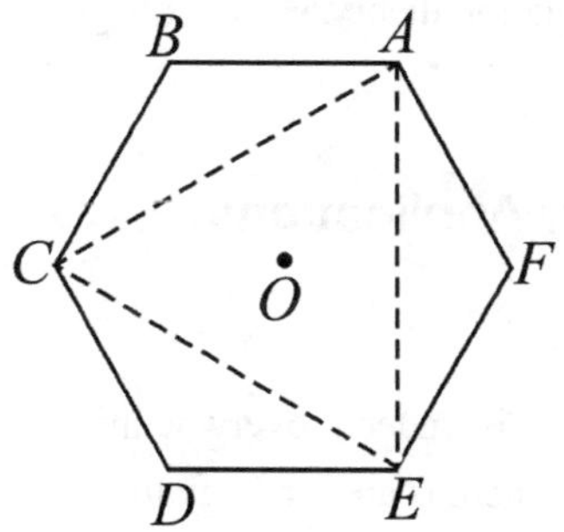

Figure 6

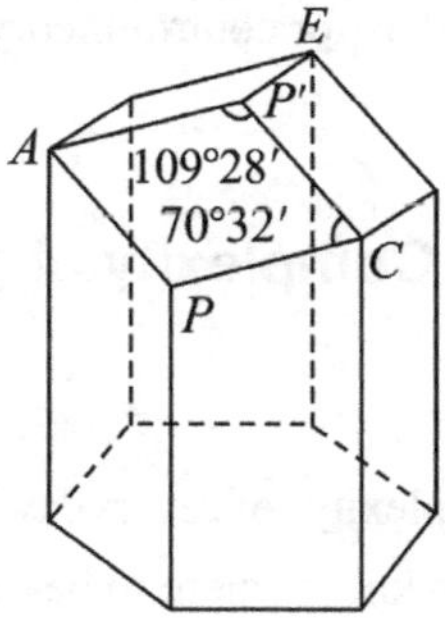

Figure 7

There is an interesting historical anecdote regarding this problem. The French mathematician and member of the Paris Academy of Sciences, König, calculated that the optimal rhombus angles for minimizing material use should be precisely 109°26′ and 70°34′. He found that the angles constructed by bees differed by only 2′. However, later, Scottish mathematician Maclaurin re-examined the calculation and discovered that the mistake was not made by the bees, but by König—who had used a logarithm table containing a typographical error. This problem, first posed in the 18th century, was one I used in 1964 to test high school students. Many of them proposed various proofs.

I have elaborated on this problem to illustrate the mathematical nature of biological mysteries. At the same time, I encourage biologists to present more problems related to biological structures. The honeycomb, for example, is closely related to crystallography—it resembles the crystal formation of a "perspective stone."

Returning to the ingenuity of chemical engineering, another mathematical problem arises: How many types of crystal structures can infinitely and seamlessly fill space? Mathematics has proven that there are only 230 possible types.

Furthermore, in the study of insulin, complex three-dimensional molecular models require intricate mathematical calculations. In genetics, the study of genetic codes—fundamental to understanding heredity and variation—inevitably involves mathematical analysis. The application of mathematics to biological reactions has contributed to control engineering and cybernetics, particularly in the concept of "feedback." Similarly, mathematical research on neural activity has provided real-world models for control theory and information theory.

■ The Complexity of Everyday Applications

The complexity of everyday applications is indeed overwhelming, making it difficult to know where to begin. However, there is an easy way to approach it—by connecting with the masses. Everyday problems affect millions of people, and by listening to the needs of workers, farmers, industries, and national defense, one can identify what should be prioritized and in what order. The people are the true heroes. Consulting with them not only enhances political awareness but also provides practical knowledge that cannot be found in books. As someone who studied mathematics through self-learning, I have gained knowledge across various fields thanks to the guidance of experts from different industries. This type of learning extends beyond what can be obtained in a university or a specialized discipline.

I first started applying mathematical methods to everyday industrial and production problems in 1958, but the real beginning was in 1964, after receiving direct instructions from Chairman Mao. I will never forget that as I took my first steps in writing *A Popular Introduction to Overall Plans* and conducting field trials, Chairman Mao further encouraged me with the directive: "Not for personal gain, but to serve the people wholeheartedly." Later, with Premier Zhou Enlai's support, I worked on *Overall Plans*. Thanks to the support of leaders from provinces, cities, and autonomous regions, I had the opportunity to travel to twenty provinces, cities, and autonomous regions, visiting thousands of factories and rural communities. I learned from millions of workers and farmers, forming small teams of workers, technicians, and mathematicians to promote and apply mathematical methods. Through large-scale scientific experimentation, we proved that mathematics is highly useful and that applying mathematical methods to innovation and efficiency improvements can create immense wealth for the nation. Looking back, it often feels like holding a golden bowl while begging for food—having valuable tools yet not fully utilizing them.

Due to the advantages of our socialist system, we may have unique approaches in this area, and I would like to share more of my experiences.

Overall planning is not limited to repairing a single machine, constructing a building, installing equipment, or completing a water conservancy project. It can also be applied to enterprise management and large-scale engineering projects. The development of the Daqing oil field required overall planning among thousands of machincs and tens of thousands of workers. In Heilongjiang, forestry operations had to run overall planning on exploitation, transportation, utilization, and reforestation. In Shanxi, coal transportation at the Kouquan Station in Datong had to conduct overall planning. The Taiyuan Railway Bureau conducted overall planning with Taiyuan Iron & Steel and several other industrial and mining enterprises, as well as agricultural production overall planning in some provinces, cities, autonomous regions, communes, and production brigades, all of which achieved excellent results. It appears that overall planning works well on a small scale but are even more effective on a larger scale. The more detailed and extensive the planning, the better the outcomes.

The key is to transfer these methods to the broader public. When combined with practical experience and implemented step by step—starting from small, simple problems and gradually expanding—it can have a significant impact. The preliminary idea can be summarized in ten words: large-scale overall planning, sorting out data, system modeling, and strategic development. The goal is to develop it into a formal discipline—Overall Planning Science—to meet China's specific needs and reflect the characteristics of our socialist society. The larger the scope of overall planning, the more data is required. But we must not merely collect data passively; we need to extract as much useful information as possible to guide decisions. This necessity brings data processing to the forefront. Handling large amounts of data efficiently requires electronic computers. Whether building new systems or upgrading existing ones, all operations must run in an optimal state. Going further, strategic development planning must be based on current conditions to determine the more proactive and reliable path for growth in the following year, ensuring the highest possible speed of national economic development.

Optimization seeks the best possible solution using the fewest number of trials. Optimization science, as the mathematical theoretical foundation for such techniques, has already undergone preliminary systematic research, and in practice, optimization has been widely promoted across many industries, including chemical engineering, electronics, metallurgy, machinery, light industry, textiles, transportation, and construction materials. Under the leadership of the Party, mass campaigns have been launched to apply optimization in all sectors and work processes. In a short time, it has been implemented in tens of thousands of experimental projects, leading to overall improvements in production levels.

Even without adding new workers, equipment, or significant or zero investments, optimization can lead to higher quality, increased productivity, and reduced resource consumption. For example, in small iron smelting furnaces, optimizing furnace shape and operating conditions improved the fuel-to-iron ratio to 1:18. In mechanical processing, optimizing tool geometry and cutting parameters doubled work efficiency. For oil-fired boilers, optimizing nozzle parameters reduced fuel consumption and eliminated black smoke emissions. In small fertilizer plants, optimization reduced coal usage while increasing output. In large chemical plants, optimizing equipment increased yields significantly. By optimizing carburetor design, a single automobile from Jiefang could save one ton of fuel per year. If this were applied to all civilian cars in China, the nation could save over 600,000 tons of fuel annually. The optimization of processing techniques in grain and rice processing can generally increase the milling yield by 1%, 2%, or 3%, and improve the flour extraction rate by 1%. When scaled to the entire national grain supply, this effectively equated to producing hundreds of millions of additional kilograms of grain per year.

The best production processes objectively exist—optimization simply provides a systematic approach to discovering them quickly and with minimal experimentation.

The rational allocation of resources, the reasonable distribution of crops, the efficient irrigation and drainage of reservoirs, the proper arrangement of power grids, and the rational layout of industries all require mathematics to solve effectively and to find optimal solutions. In short, it is a mathematical problem to seek the most reasonable solution (according to a specific goal, such as the most economical or the least labor-intensive) in a system where various factors are interdependent and mutually influential. This is the concrete embodiment of the "more, faster, better, and cheaper" principle. There are many mathematical methods used in this process, and we have prepared some that are applicable. However, due to the interference and destruction caused by Lin Biao and the "Gang of Four," there has been no capacity for in-depth work. Today, as we open a new chapter in socialist construction, mathematical research and application will also usher in a brand-new situation.

■ The Development of Mathematics

The vast universe, the tiny particles, the speed of rockets, the ingenuity of chemical engineering, the changes of earth, the mysteries of biology, and the complexity of daily applications, mathematics is indispensable. For example, Einstein used mathematical tools to derive formulas that pointed to directions for finding new energy sources and predicted the energy released during nuclear fission. Even abstract concepts like fiber bundles are applied in physics. In astronomy, calculations first predicted the existence of Neptune, which was later discovered. Similarly, in high-speed flight, the transition from subsonic to supersonic speeds led to the study of hybrid partial differential equations. Other fields, such as the relationship between radio electronics, computational technology, and information theory, the connection between automation and control technology with ordinary differential equations, the relation between the nervous system and cybernetics, and the connection between morphogenesis and structural stability, all involve mathematics.

Mathematics is a field rich in generalizations. Abstraction is its hallmark. Take, for example, the same equation: In elasticity mechanics, it describes vibrations; in fluid mechanics, it models fluid dynamics; acousticians may call it an acoustic equation, while electrical engineers may refer to it as a telegraph equation. What mathematicians study is the common aspect of these phenomena—the hyperbolic partial differential equation. The properties of this partial differential equation are the shared characteristics of these different objects, and its numerical solutions are the data required across the various disciplines it connects.

Moreover, these shared properties can foster the possibility of unifying theories across different branches and enable the mutual simulation of different phenomena. For instance, acousticians can use similar electrical circuits to study acoustic phenomena, significantly simplifying the complexity of acoustic experiments. The most common application of such simulation is the development of electronic computers. Given that nerve cells have two states, excitation and inhibition, and that electrical systems have charged and uncharged states, mathematics, using binary digits (0 and 1), and logic, employing "yes" and "no," has led to the creation of electronic digital computers that attempt to simulate the nervous system and the early results of simulating logical thinking.

Our explanation above does not suggest that mathematicians should immerse themselves in studying only these shared properties. Admittedly, we must acknowledge that requiring mathematicians to delve into every aspect connected to their research is highly challenging. However, this does not exclude the necessity for mathematicians to deeply explore the various fields of science they are connected to, or even parts of those fields. This kind of in-depth engagement is entirely essential. It not only contributes to national economic construction but also greatly benefits the development of mathematics itself.

In general, the emergence of objective phenomena falls into two main categories: One is the phenomenon of necessity—also known as the law of causality, and the other is the phenomenon of large numbers—also known as the law of chance. The mathematical tools that describe phenomenon of necessity are generally equations. These equations can derive unknown data from known data and deduce the properties of unknown phenomena from known phenomena. Common examples include algebraic equations, differential equations, integral equations, difference equations, and particularly, differential equations. The mathematical tools used to handle phenomena involving large numbers are probability theory and mathematical statistics. Through such analysis, we can understand the general trends and the proportional patterns of different situations.

Other branches of mathematics can certainly be directly linked to real-world problems. For instance, mathematical logic is related to the design of computers and automation, complex function theory is applied in fluid mechanics, functional analysis and group representation theory are used in quantum mechanics, and Riemann geometry connects to relativity theory, among other applications. Number theory is also employed in computer design. Generally speaking, mathematics itself is an organic whole, interconnected, and the areas mentioned above are among the most frequently and widely connected with other sciences.

Computational mathematics is a discipline that has existed since the beginning of mathematics, but today it has gained particular significance due to the advent of fast, large-scale computers. As the complexity and scope of problems increase (the computational workload for a single problem has reached unprecedented levels), the importance of computational mathematics has become even more apparent. For example, solving a system of equations with 100 unknowns is a common task in science today (such as stress analysis for dams, geodesy, designing suspension bridges, large-scale architecture, etc.), and solving it manually is very difficult. Similarly, calculating weather equations to predict tomorrow's weather based on today's data would take years if done manually. So, what does an algorithm have to do with tomorrow's weather? It's a mere irony! The invention of the electronic computer met this demand. The development of high-speed, large-storage computers meets this requirement, but the advent of modern electronic computers has not diminished the importance of mathematics. On the contrary, it has enhanced the power of mathematics and raised the demands on it even higher. While the burden of heavy computational labor has been lightened or eliminated, the need for creative labor has increased. Computational mathematics acts as a bridge, linking the creativity of mathematics with practical applications. At the same time, it is a creative discipline in itself, for instance, driving the development of a new field, computational physics.

Apart from the branches particularly emphasized above, this does not mean that other areas of mathematics are unimportant. It is only through the close interconnection of these key areas with others, and the cooperative development of pure and applied mathematics, that we can meet both the current and long-term needs of our country.

From the historical development of mathematics, we can observe that as society advances, the scope of applied mathematics expands, the mathematics applied becomes more sophisticated, and the number of people involved in applied mathematics increases. In ancient societies where people worked from dawn to dusk, basic counting sufficed to meet practical needs. Later, with the need to track the seasons and measure fields, geometry emerged to study the world we live in. As commerce grew, more calculations were needed, leading to the development of algebra. To describe dynamic phenomena and study changes in relationships, analytic geometry and calculus emerged.

The usefulness of mathematics in the physical sciences has already been historically proven. Its role in biological and social sciences has also begun to show, with vast potential ahead.

Finally, I must state that I am not suggesting that other sciences are unimportant or of lesser importance. What should be emphasized is that mathematics is important precisely because of the importance of other sciences. Without other disciplines, the power of mathematics could not be demonstrated, nor could it be deemed important.

It should be pointed out that the "Gang of Four," in their attempts to restore capitalism, ruthlessly destroyed production and hindered the development of science and technology. They not only sabotaged theoretical research but also launched brutal attacks on those engaged in applied mathematics. The harmful effects of their actions must be thoroughly eradicated and should not be underestimated. In order to realize the "Four Modernizations" and build China into a strong socialist nation, the importance of developing mathematics is indisputable.

Study, Think, and Persevere*

Recently, young students have responded to the Party's call to strengthen their studies, creating a more intense academic atmosphere in schools than before. This is a very encouraging. Here, I would like to offer a few preliminary thoughts on studying for discussion.

■ Learn to Self-Study

From elementary school to junior high school and then to university, students have been studying for more than ten years. Yet, I still emphasize the importance of learning how to study. This may sound strange, but in fact, it is not. Learning how to study is not simple, and I, too, am constantly exploring and improving my methods. One must not assume that memorizing texts fluently means truly understanding them. If one does not gradually deepen comprehension, what is the difference between that and a monk mindlessly chanting scriptures? I believe that learning how to study is just as important as acquiring specialized knowledge during school. Mastering the ability to study ensures not only good academic performance in school but also continuous self-improvement in the future.

The time we spend working in life is much longer than the time we spend in school. A young person who may not have graduated from university or even high school, but who has developed the habit of self-study, can still achieve success in their career as great as college graduates. On the other hand, if a person earns a university degree, studies abroad, learns from prestigious mentors, and even obtains a doctorate, yet has not learned to study and explore independently, they will remain

---

* This article went through supplement and revision based on the author's speech at the opening ceremony of the University of Science and Technology of China. It was first published in issue 21 of *China Youth* in 1961.

confined within the boundaries set by their teachers, unable to expand their knowledge, let alone make original contributions to scientific research.

So how should we learn to study? I believe, when studying a topic or a chapter in a book, one should not only focus on the text itself but also seek to understand the ideas behind it. That is, when learning about certain principles, laws, or formulas in a textbook, it is not enough to memorize the conclusions or understand the reasoning. One should also ask: How did the author arrive at this conclusion? What challenges did they overcome? Furthermore, imagine if the book did not provide the answer—how would you derive it yourself?

Engels once said, "What we need is not just the bare result, but the research process itself. If we fail to understand the development leading to a conclusion, then we have not truly grasped the result." Only by understanding how a conclusion is derived can we truly grasp it. Only by knowing not only what something is but also why it is so can one gain a thorough understanding of the problem. To achieve this, we must study each problem in the book until we truly understand it—if we don't grasp it in one day, we should continue studying it the next day. If we haven't understood one chapter, we shouldn't rush to the next. Although this approach may be slower, it yields real results.

When I was young, I often rushed through books, quickly finishing them. Initially, this seemed like progress, but when I needed to apply my knowledge, I realized I couldn't—I had only skimmed the surface. Fortunately, I had only a few books at the time, so I was forced to return and study them more thoroughly (not merely repeating the reading but deepening my understanding), which significantly benefited me.

If the first step in studying is analysis—breaking down the material—then the second step is synthesis—connecting different parts. After thoroughly understanding each topic, one must link different sections together, gaining a comprehensive grasp of the book's core themes and the relationships between its ideas. This allows us to identify the key principles that run through the book and truly internalize its essence.

I often refer to this process as "from thick to thin." Before reading a book, it may seem lengthy and daunting. As we read and annotate every chapter, supplementing with reference materials, it may feel even more overwhelming. However, once we deeply understand the book's content, pinpoint its key ideas, and grasp its essence, the book suddenly seems much thinner. The more thoroughly one understands, the lighter the book feels. Every scientist experiences this process. This does not mean we have learned less—rather, we have fully digested the material.

Young students should learn to digest knowledge. I often see students asking teachers to highlight key points before exams, which shows that they have not yet grasped the main ideas themselves. Relying on teachers to point out key points is not an effective approach—one must develop the ability to identify them independently.

When reading a book, we should compare it with the knowledge we have previously learned and think about what new things it adds. When reading a new book, we can quickly go through the parts we already understand, but we should focus on studying the key points in depth. We should devote even more effort to new concepts. After finishing a book, the goal is not to memorize it entirely but to gain a few new methods or insights previously unknown to us. This approach leaves a deeper impression and strengthens memory. Moreover, the more we learn and understand, the stronger our knowledge foundation becomes, allowing us to read and comprehend at a much faster pace.

■ Learn to Think Independently

The reading methods discussed earlier are essentially about developing independent thinking skills during the learning process. Our endeavors are constantly advancing, and after graduation, regardless of the field one enters, one will inevitably encounter numerous new problems. In our lifetime, it is rare to find ready-made answers in books for the challenges we face. More often, we must fully utilize our ability to think independently and apply knowledge flexibly to solve new problems. For those engaged in scientific research, the very purpose of their work is to uncover what was previously unknown. Every major scientific invention in history is the result of deep, independent research on the problems by its discoverer. Therefore, young students should cultivate the habit of independent thinking while still in school.

To think independently means, on the one hand, to inherit the achievements of predecessors, but on the other hand, not to be constrained by them. If one disregards past accomplishments and blindly explores without direction, they are bound to take unnecessary detours. It is possible to spend a great deal of time researching something, believing it to be a new discovery, only to find out that it was already known decades ago—resulting in wasted effort. Sometimes, rejecting the achievements of predecessors may sometimes lead us into a dead end from which we cannot escape, such as attempting to trisect an arbitrary angle with a compass and straightedge in mathematics or creating a perpetual motion machine in physics, both of which have long been proven impossible. Even the most innovative scientists in history were those who best absorbed their predecessors' achievements. Newton famously said that his great scientific accomplishments were possible because he stood on the shoulders of giants, building upon the foundations laid by those before him. In our daily lives, we also observe that young students who are willing to learn humbly tend to progress much faster than others.

However, accepting past achievements can sometimes constrain one's thinking. To truly think independently, one must not only absorb the knowledge of predecessors but also break free from their limitations. This allows us to raise and solve problems that others have not considered and to refine, expand, or even discard previous conclusions and methods. In fact, surpassing our predecessors requires independent thinking. Today, if we aim to catch up with and surpass advanced countries in scientific research, we must have an innovative spirit. If we merely follow in others' footsteps without exploring new paths, we will always lag behind. To surpass others, originality is a must.

Independent thinking requires both bold imagination and a commitment to truth. Scientists must dare to challenge conventions, break free from outdated superstitions, and explore ideas freely. Without the courage to think beyond established norms, how can new inventions and discoveries emerge? However, scientific imagination must be combined with a rigorous commitment to reality. For instance, the romantic idea of space travel has existed in Chinese literature, poetry, and myths for centuries. Yet it was only a little over a hundred years ago that Russian scientist Konstantin Tsiolkovsky proposed using rockets for spaceflight. It took many more years of diligent research to solve the technical challenges of rocket propulsion before this dream became a reality. In scientific research, it is not only valuable to have bold ideas but also to steadily work toward making them a reality.

To develop independent thinking skills, we must consciously train ourselves to do so. We must be willing to engage in deep thought and regularly question the things we encounter. For example, if we read in the news that the Soviet Union plans to launch a rocket into the Pacific, as students of mathematics, we might calculate the launch site, range, and accuracy based on the four published launch zone coordinates. Frequently engaging in such exercises, even if they do not seem immediately useful, helps develop the habit of analytical thinking. Scientific discoveries are the result of persistent intellectual effort accumulated over time. Every major breakthrough is built upon a series of smaller, seemingly simple steps. Whether we regularly engage in deep thought will become evident when tackling major research challenges. Solving significant scientific problems always requires overcoming numerous difficulties. However, for those accustomed to thinking regularly, many of these difficulties will have already been considered in advance—leaving only a few major hurdles to overcome. Such individuals can conduct research more efficiently and see further, think deeper, and understand more thoroughly than others.

The reasons that hinder us from thinking regularly can be summarized into two: fear of difficulty and underestimating problems. Those who fear difficulty find a problem daunting even before attempting to think about it, so they naturally avoid engaging with it. Meanwhile, those who see problems as too simple consider them unworthy of deep thought, thereby missing opportunities for in-depth research and new discoveries. In reality, many problems may appear simple and easy at first glance, but a deeper examination often reveals their complexity. Even if a problem is indeed simple, taking the time to think about it can sometimes lead to new insights. Let me give a common example from daily life: Suppose a family of nine people consumes half a tael of oil per person per day. How much oil does the entire family consume in a month? This problem is easy, even for elementary school students. Most people would calculate it as $(9 \times 0.5 \times 30) \div 16 = 8$ *jin* and 7 taels (1 *jin* = 10 taels, 1tael = 50 g). However, upon further thought, one might realize a more efficient way to calculate it: Since 30 days of consuming half a tael per day equals 1 *jin* minus 1 tael, for nine people, it would be 9 *jin* minus 9 taels. Isn't this method much simpler? This demonstrates that we should not overlook valuable insights by making our mental filters too coarse, or we may miss out on important discoveries.

■ Perseverance Is Essential

The reason why we need perseverance is that any major achievement in scientific research requires enduring dozens, hundreds, or even thousands of failures before success is reached. For a scientist, failure is far more frequent than success, and their experience with failure is significantly richer than their experience with success. Some young people, upon seeing renowned scientists publish valuable academic papers, assume that these individuals must be geniuses who effortlessly produce important work. What they fail to realize is that the reports they see only showcase the successful portion of the scientists' research. The numerous failures that preceded those successes remain unseen. The process of scientific research is an upward struggle, often filled with setbacks. At times, success seems within reach, only to be lost again. Conversely, moments of failure can, through persistent effort and deep contemplation, lead to new breakthroughs, as described in the Chinese saying, "Beyond the dark willows and bright flowers lies another village." As Karl Marx famously said, "There is no royal road to science, and only those who do not dread the climb up its steep paths have a chance of reaching its luminous summit."

Modern scientific development is characterized by two key trends: increasing specialization and rapid growth in interdisciplinary fields, where the connections between different sciences are becoming closer. This complexity in modern science can only be overcome through long-term research. Sporadic efforts with long gaps in between will not lead to success. Some young people seek to take on only easy scientific problems, but in reality, most simple questions have already been addressed by past generations. The challenges we face today are far more complex. At the same time, thanks to the diligent research of our predecessors, many foundational problems have already been solved, providing us with a strong basis for further exploration. Because of this advantage, we should set even higher expectations for ourselves.

In scientific research, one must avoid relying on luck. No scientific discovery occurs by mere chance. Some young people believe that Newton discovered the law of universal gravitation simply because he happened to see an apple fall from a tree. In truth, Newton's discovery was not merely a result of witnessing the falling apple—after all, humans had observed apples falling for millennia. Rather, it was his extensive study of Kepler's laws of planetary motion and Galileo's laws of falling bodies, coupled with long-term contemplation, that enabled him to derive the principle of universal gravitation when he saw the apple fall. Scientific inspiration is not something that can be passively awaited. If there is any element of chance in scientific discoveries, it is only available to those who are well-versed in their field, capable of independent thinking, and possess unwavering perseverance. It will never come to those who are lazy.

■ Great Ambition and Meticulous Planning

Our country is currently engaged in a grand socialist construction. Young people are the driving force behind this effort and the successors of the older generation of revolutionaries. In scientific research, it is essential to cultivate the ambition to climb the peaks of science. Those without great ambition lack the drive for a remarkable life, and naturally, they cannot be expected to achieve outstanding accomplishments.

However, once ambition is set, it must be accompanied by a well-thought-out plan to realize it. We must carefully plan the steps to achieve our great ambitions, ensuring that our efforts bring us closer to our goals step by step. If our target is in the east but we head southeast, we will waste our energy and end up working against our own intentions. On the path forward, we will inevitably face many obstacles, and the further we progress, the more difficult they become to overcome. We must tackle them systematically, one by one. If we have not yet conquered the first challenge, we should not attempt to take on the second. Rome isn't built in one day. In scientific research, those who are too eager for quick success often end up moving the slowest. We must strive to advance both quickly and steadily.

Young students, you are fortunate to live in the era of Mao Zedong, where the state has provided excellent conditions for your education. Whether you succeed in your studies depends on your diligence and commitment to learning. The Party and Chairman Mao have already given comprehensive guidance to help you learn well. What I have written here is merely my personal experience, shared for your reference.

Thrilled to See Young Saplings Thrive*

■ Exciting News Brings Joy

Late at night, when everything quiets down, the ringing of the telephone sounds particularly crisp. Who could it be? A bit unexpected—no, actually, it was expected. This was the call I had been subconsciously waiting for all evening. Even before picking up, I had an idea of what the news might be. But no matter how much one estimates, reality often surpasses expectations—the news was even better and more exciting than anticipated.

"I have great news for you. The grading is now complete, and the results of the senior high school mathematics competition are unprecedentedly good. Everyone has passed the 'Labor and Hygiene Standard,' and many have even exceeded the 'Champion Standard.'" This was how Comrade Gong Sheng, Secretary-General of the Beijing Mathematical Society, humorously delivered the message over the phone. The humor comes from this: "This morning, two rounds of the mathematics competition unfolded. The first round tested basic skills and fundamental training. The results? Everyone passed, with some even achieving full marks. The second round assessed the flexible application of mathematical abilities, and it truly revealed the students' levels—especially the senior high school students, who achieved an unprecedented standard. Even under the strict evaluation of the judges, the top scorer reached an impressive 86 points."

This was truly fantastic news! Compared to previous years, there was a significant improvement—how could one not feel happy, encouraged, and excited?

---

* First published in *Beijing Daily* on May 7, 1962.

■ Hard Work Bears Fruit

Excitement made it difficult to fall asleep. In my mind, I saw lush green fields filled with robust young crops, hinting at a bountiful harvest ahead. This reminded me of the scene on the morning of the competition:

The competition was set to start at 8 a.m., but many students arrived early. Some teachers accompanied them, and even some parents personally escorted their children. Their emotions mirrored those of the competing students—joyful, excited, yet nervous. The students, eager and determined, were ready for the challenge.

As the bell rang, the competition began. The tension was equally intense both inside and outside the venue. Students focused, pencils moving swiftly, striving not just to solve the problems but to do so quickly, accurately, and elegantly. Outside, teachers were also given copies of the questions, attempting to predict their students' performance based on the difficulty of the test—much like farmers estimating their yield. They concentrated on solving the problems, preparing to help students analyze their results after the first round.

One parent even traveled twenty li to bring a compass to their child, who had forgotten to bring one in the rush of the morning.

After the first round ended, students and teachers immediately engaged in discussions. Students formed small groups, calculating aloud, drawing on the ground, and analyzing their performance. Teachers gave careful advice: "The first round went well, but the second requires even more composure and attention to detail. Don't rush to submit your paper once you've finished—review it carefully!" This level of meticulous care and encouragement from teachers was evident not just today but throughout the years, through every bit of effort they invested in their students.

It reminded me of the past three months of hard work by the exam committee members. Designing problems is much harder than solving them—they must be clever, well-structured, and capable of distinguishing different levels of ability. Committee members spent countless hours brainstorming, drawing from their own scientific research experiences. Once the problems were set, they had to be reviewed one by one, discussed, refined, and finalized. Standard solutions were also prepared as reference materials for teachers.

Even today, the committee members spent the entire day supervising the competition and grading the exams, only leaving after night had already fallen. Meanwhile, members of the Beijing Science Association and other staff had spent the past month tirelessly printing exam papers, preparing materials, and organizing venues.

■ Fertile Soil and Warm Sunshine

Why does everyone work so hard? It is for socialism, for the future of our nation's youth. Seeing young saplings grow strong, I couldn't help but think of the diligent efforts of farmers—sweat dripping into the soil, toiling tirelessly. Similarly, when we see such outstanding competition results, we must acknowledge the immense contributions of teachers, who have laid a solid foundation for the country's scientific future. Thanks to them, we can envision an even brighter future for Chinese science.

While we express our gratitude for the hardworking farmers, we must also not forget our proud, excellent environment—abundant sunshine, fertile soil, and favorable weather conditions.

Today's sunshine is the light of Chairman Mao, the radiance of our beloved Party. The soil in which these young talents grow is socialist soil—the most suitable environment for scientific advancement. The Party's education is like the spring breeze and nourishing rain. In such a supportive environment, we have teachers who care deeply for their students, and we see strong, thriving young saplings. This harmonious combination of favorable conditions—the right time, the right place, and the right people—creates an ideal setting for progress.

■ Beyond the Mountains, Beyond the Skies

This is not a 100-meter sprint; it is a marathon. No, even calling it a marathon is insufficient—it is far longer. The pursuit of scientific achievement is an endless journey, one that takes a lifetime, with the torch being passed to future generations. Of course, along the way, there will be moments of great success—brilliant peaks reached one after another. But beyond every mountain, there is another mountain; beyond every sky, another sky. Those with greater perseverance will go further, climb higher, and make even greater contributions to the people and to socialist development.

Throughout history, all great scientists have shared one trait—humility. This humility is not necessarily something they were born with; rather, the more they learn and think, the more they realize how much they still do not know and how many more problems lie ahead for them to study. They will also realize that while there are problems they can solve that others cannot, there are even more things that others can think of that they cannot. Therefore, they come to have a correct self-assessment and understand the truth that humility leads to progress.

At any time, even a little arrogance and complacency can slow us down. The happiest people are those who are always aware of their own shortcomings. Recognizing our deficiencies drives us to work harder, to strive for excellence. To the winners of the competition: Do not let even the slightest trace of arrogance creep in—pride breeds laziness, and laziness leads to falling behind. Conversely, recognizing one's shortcomings fosters diligence, and diligence leads to success. A fast start does not guarantee an early finish—persistence is the key to victory. Likewise, those who did not achieve outstanding results in the competition should not feel discouraged. Through continuous effort, we can surpass one milestone after another, always reaching greater heights.

■ Practice Makes perfect

Practice makes perfect. Basic skills must be practiced frequently, whenever there's an opportunity. In daily newspapers, we often encounter various problems that can serve as material for our practice. For instance, one can analyze how the Soviet Union's first artificial satellite will pass by earth again in the year 2113. Similarly, by examining the restricted landing zones for Pacific rocket launches, one can infer the possible launch sites. Examples like these are everywhere.

With constant attention and practice, one will continuously improve. When faced with a challenging problem, do not give up easily. Do not dismiss it as something only rocket experts should think about. Instead, reflect on it, discuss it. If you cannot solve it today, you may be able to solve it in the future when you have gained more knowledge. (In fact, the two examples mentioned above can already be solved—or at least learned to be solved—by senior high school students.)

Of course, not everyone can participate in a math competition. However, those who never do, through hard work and diligent study, will also scale one peak after another. May these young saplings continue to grow strong and flourish!

Dream Bigger*

The same event can be reflected in different ways. Those who strive upward focus on the positive aspects, while those who give up focus on the negative.

In the Song Dynasty (960–1279), there was a man named Su Xun (also known as Su Laoquan), who only began studying seriously at the age of 27, eventually becoming a renowned literary figure.

This story teaches us to study hard and don't be discouraged by the notion that it's too late to start. Don't say that at my age, I won't achieve anything significant. Nor should you think, "I'm still young, just 20 years old, why should I put in effort now? Isn't it better to start later, just like Su Laoquan did, two years later?"

It's better to start early. If you start late, don't regret it. The most important thing is to avoid not trying at all, and to not live your life in ignorance.

Thinking about how our impoverished country is waiting for us to build it, about the difficult and complex nature of building socialism and communism, about the hopes of Chairman Mao and the Party, we are eager to master skills immediately. We must not slack off or waste time, and should resolve to act early and avoid hesitation.

During the Yuan Dynasty, there was Wang Mian (also known as Wang Yuanzhang), a great painter. He was born into poverty and could not afford to study art with a teacher. Instead, he bought paper, brushes, and paint, and while tending to cows, he observed lotus flowers and practiced painting them. After years of hard work, he became a great artist.

This story tells us: If there's no teacher, don't be afraid. With diligent self-study and effort, you can still achieve great things. However, this does not mean we should disregard formal education and rely only on self-study.

The habit of self-study is valuable and beneficial throughout life, but don't be unaware of the blessings you have! Don't fail to recognize the value of the help

---

* First published on June 16, 1962, in *China Youth Daily*.

provided by teachers. In areas that are unclear, teachers explain; in areas that are deep, teachers investigate. More importantly, from teachers, you can learn the lesson of both hard-earned successes and frustrating failures. By learning from their successes, you gain knowledge and can delve deeper into the subject. By learning from their failures, you gain experience and understand that talent comes from hard work, and success arises from overcoming failure.

The spirit of originality is essential. This is a skill that young people must possess in this era, as we work to build something unprecedented. But accepting the achievements of those who came before us is equally important. Avoiding or minimizing the mistakes that others have already made is a good way to speed up progress. Teachers give their best effort to guide us to the highest level they can. By standing on their shoulders, we can rise even higher, one generation surpassing the last.

During the Five Dynasties (907–960) period, there was a poet named Wei Zhuang (Wei Duanji), who wrote the following *ci* (a type of lyric poetry):

Street drums beat,
The Forbidden City opens,
A person returns from the heavens,
A phoenix carries the golden list out of the clouds,
A thunderous sound on the ground.
The oriole has migrated,
The dragon has transformed,
In one night, the city is filled with carriages and horses,
Every family on their rooftops worships the immortals,
Competing to watch the crane soar into the sky.

This *ci* likely reflects the situation of candidates succeeding in the imperial examination system, but it also includes the poet's romantic fantasies—his rich imagination.

But fantasy is still just fantasy. To truly reach the heavens and return, to truly bring scientific data "out of the clouds," it still requires the hard work of scientists over thousands of years.

Ambition is essential, and while romantic fantasies are also important, hard work, perseverance, and systematic progression are even more important. Be honest, practical, and never underestimate the value of taking steady, gradual steps. Every step forward brings us closer to our goals, and the higher we climb, the more we must remain humble. Inspired by Wei Zhuang's *ci*, we may dream of immediately inventing a spacecraft and soaring through the universe, but that is a pipe dream. Only through gradual steps—first thinking through the possibilities, calculating data, and analyzing the requirements systematically—can we achieve today's accomplishments.

Time has unfolded over a century, and countless wisdom has been poured into it. In the grand sweep of history, our work may seem but a tiny ripple, and in the vast expanse of this era, it may appear insignificant. Yet, it is through these small steps and humble contributions that great discoveries and innovations emerge. Like drops of water gathering to form a river, each little effort accumulates, propelling us toward greater heights. With a vision that stretches far and a pace that remains steady, we must avoid lofty ambitions that outstrip our abilities. True greatness is born not from boasting, but from diligent, consistent work. Those who boast of their achievements often find their paths fraught with obstacles, while those who toil quietly, step by step, are the ones who create lasting impact.

Talking to Youth About Learning*

Young students often hope that I will talk to everyone about learning. Although I am a little older than most young people, I am still learning through exploration, constantly gaining lessons from my failures. I do not yet have a set of mature experiences or perfect methods for learning, but I do have one aspiration: to learn for a lifetime, and to never lose heart, never allow temporary setbacks to diminish my enthusiasm and resolve for learning.

Science is a field built on sincerity, where even the smallest trace of falseness cannot be tolerated. Therefore, I will first sincerely lay out some thoughts before moving on to the main subject. What I discuss today is meant to be thought over by everyone—to see what is right, what is wrong, what can be absorbed and refined, and what should be discarded. Perhaps, in this way, one may gain something useful from others while avoiding their mistakes.

■ Be Ambitious

We are now facing a great era. Our task is to transform our impoverished and backward country into a new China with modern industry, modern agriculture, and modern scientific and cultural achievements. Shouldering such a task, every young person should set great ambitions, with the vitality to fight, to succeed, to learn, to innovate, to continue the work of those who came before us, and to take on some unprecedented monumental projects.

---

* First published on December 8, 1962, in *Yangcheng Evening News*.

However, this does not mean that only those involved in cutting-edge scientific research or innovation need ambition and those with more mundane work do not. Even the most routine work can be perfected. As the saying goes, "Every profession has its master." Without ambition and determination, one cannot proactively excel in their work. Some people think that working in agriculture doesn't require ambition, but that is incorrect. If someone's efforts can increase agricultural output by just 1 percent, that is already an extraordinary achievement. To do that, one must have ambition. Others believe that teaching in primary and secondary schools does not require ambition, but this is also wrong. Cultivating the next generation is one of the most fundamental tasks in national development. To think that primary and secondary school teachers don't need to study hard and research their subjects is a discouraging attitude, not one of striving for excellence.

A primary school teacher in Beijing once said, "What elementary school students need might be just a glass of water, but we must prepare a full pot of water to meet their needs." This is very true. To be a truly excellent socialist teacher, one must accumulate knowledge and be ready to satisfy the thirst of the young. My own experience is similar. Sometimes, when I teach a first-year university math course, which is only one year ahead of senior high school's, I feel that my knowledge is not enough—far from it! It's always lacking. I often discover new, better materials or teaching methods. I constantly feel that there is room for significant improvement in my teaching. Even after preparing a lecture well, I realize after delivering it that there are many areas for improvement. The ancient wisdom of "teaching benefits both the teacher and the student" couldn't be truer.

I believe that in any job, one should never be satisfied with mediocrity or simply go through the motions. Instead, we should strive for excellence and continually improve.

■ To Reach the Skies, One Must Start from the Ground

Ambition must be accompanied by a solid and methodical approach to learning. True ambition is not about being overly eager or rushing for quick success. Unlike empty fantasies, real ambition is grounded in careful planning—laying out concrete steps to achieve one's goals in a steady and practical manner.

For instance, who wouldn't want to soar into the skies? But mythical tales like Chang'e flying to the moon or Sun Wukong somersaulting across the heavens are just fantasies. To truly reach the sky, one must start from the ground. The poet Wei Zhuang of the Five Dynasties wrote two versions of "Xiqianying" (Joyous Migrating Orioles):

The crowd stirs, the drums roll,
Chilled by the winds before dawn.
Above the grand heavens, the moon hangs hazy,
Astride a celestial steed, riding into the void.
Fragrance fills my robe, clouds line my path,
Phoenixes circle and dance around me.
Bannered processions stream in flocks,
Led forward to meet the Jade Flower Lord.

Street drums beat,
The Forbidden City opens,
A person returns from the heavens.
A phoenix carries the golden list out of the clouds,
A thunderous sound on the ground.
The oriole has migrated,
The dragon has transformed,
In one night, the city is filled with carriages and horses.
Every family on their rooftops worships the immortals,
Competing to watch the crane soar into the sky.

These are poetic dreams—beautiful, yet detached from reality. To truly explore the skies and bring back scientific discoveries from above, it takes generations of dedicated effort.

Some students in junior high school attempt to self-study quantum mechanics—is this ambition? Perhaps, but it may be premature. Without understanding classical mechanics or calculus, it is impossible to grasp quantum mechanics. Broadening one's knowledge is important, but one must also be mindful of their current level. Learning must be solid—each step built on the last. Skipping steps leads to setbacks, and repeated missteps can sap confidence entirely. However, if you ever find yourself missing a step in your learning, don't be afraid—there is always a way back. Seize the opportunity to make up for it without embarrassment. If you don't fill the gap, it will remain a void forever; but once you do, the issue is resolved, a weakness is corrected, and your progress will become steadier, firmer, and faster.

■ The Requirement of Understanding

A strong foundation in learning makes it easier to reach new heights, master complex subjects, and apply knowledge effectively. Some believe a broader foundation is better, but how broad is enough? Some attempt to read an overwhelming number of books to build their base. I believe the first requirement for building a strong foundation is to thoroughly understand the fundamental concepts. One should not rush into reading books beyond their current level of comprehension. What does it mean to truly understand something? It involves a process of "from thin to thick" and then "from thick to thin."

From thin to thick: When studying a book, you must look up every unfamiliar word, analyze every unclear sentence, and annotate confusing sections. After putting in this effort, you may feel that you understand much more, yet at the same time, the book seems to have become even "thicker." Some may think that this means they have fully understood the material. However, that is not the case. Even if every chapter, every section, every word, and every sentence is understood, it is still not the final form of understanding.

From thick to thin: The final step is the process of "going from thick to thin." This requires thoroughly digesting what has been learned—analyzing, organizing, and refining it through repeated contemplation and integration. One must extract the key concepts, trace the logical flow, identify the main points, and compare them with previously acquired knowledge to understand what new ideas and methods have been added. Through this process, the once dense material begins to feel lighter and more concise. Knowledge that has been deeply absorbed in this way becomes easier to remember and can be applied with confidence and ease.

For example, in mathematics, merely memorizing formulas is ineffective. True understanding comes from grasping the underlying principles. Trigonometry contains numerous formulas, but most derive from a few core ones. One of them:

$\sin^2\theta + \cos^2\theta = 1$ is not even new—it originates from the Pythagorean theorem in geometry.

■ Learning Accelerates with Mastery

Some may worry that delving deeply into a subject might hinder breadth of knowledge, or that such an approach might slow down progress. However, the opposite is true—this method actually leads to faster learning in the long run. A strong foundation means that future learning involves simply adding new knowledge, rather than cramming an entire book into one's brain each time. Instead of indiscriminately accumulating information, this approach transforms "breadth" into "enhancement"—adding what is truly unknown rather than just piling up more of the same. This deep and broad digestive learning method is both effective and efficient, ensuring that knowledge, once acquired, remains firm and readily accessible.

Another key to building a solid foundation is constant practice—seizing every possible opportunity to practice what has been learned. For example, when studying mathematics, one should not be content with merely solving assigned problems but should challenge themselves with exercises from other schools or even math competition problems. Reading a newspaper and seeing a report about a five-year plan requiring to double a product's output? Why not calculate the annual percentage increase? Learning about the launch coordinates of a ballistic missile? A student of mathematics might explore what insights can be derived from that data.

■ Beyond the Classroom

Alongside building a strong foundation, it is crucial to develop independent thinking. The world is constantly evolving, and old methods may not always be suitable for new problems. When faced with unprecedented challenges, one must carve out new paths and devise original solutions. What the teacher has not explained, what the textbooks do not cover, and what past scholars have never encountered—all these demand independent thinking.

The first step toward independent thinking is still rooted in a solid foundation: extensive exercise, persistent intellectual effort, and a thorough understanding of the origins and implications of key theorems, laws, and formulas. However, it is not enough to merely comprehend these principles—one must also ask that how did others arrive at these conclusions? Once one understands the thought process behind past discoveries, they are more likely to generate their own new insights.

■ Was Newton's Discovery Just Luck?

Emphasizing independent thinking does not mean disregarding the work of those who came before us—on the contrary, true innovation is built on a deep appreciation of past achievements. Some in capitalist societies have popularized the myth that scientific breakthroughs occur purely by chance, as if great discoveries are stumbled upon through luck. They claim that Newton formulated the law of universal gravitation simply because he happened to see an apple fall from a tree, as if the insight came in a sudden flash of inspiration. This is utter nonsense! People have watched apples fall since the dawn of humanity, so why was it Newton who uncovered the law of gravitation?

In reality, Newton's breakthrough was not a mere coincidence. He had spent years studying Kepler's laws of planetary motion and Galileo's principles of falling bodies. His discovery was the result of long-term learning, deep contemplation, and rigorous analysis. When he encountered a familiar natural phenomenon, his prepared mind allowed him to perceive its underlying essence. Scientific progress is a methodical pursuit—it does not rely on luck. Without a deep understanding of a problem's core principles, even a serendipitous encounter with a key phenomenon would be meaningless. Many people walk into a treasure trove only to leave empty-handed, simply because they do not recognize the value of what lies before them.

We must humbly learn from others' successes, but just as importantly, we should pay attention to their failures. Observing how great minds faced difficulties and overcame setbacks often provides even more valuable lessons than studying their achievements. It is easy to admire the books published by scholars or the articles they contribute to newspapers and journals, but few realize that for every published work, countless discarded drafts fill their wastebaskets. Understanding the immense struggles they endured in their pursuit of knowledge allows us to learn from their perseverance and problem-solving strategies.

■ Diligence Overcomes Inadequacy, Practice Makes Perfect

Finally, let's address the relationship between talent and learning. Some people lack confidence, believing that success in learning requires talent—something they feel they lack. Others, on the contrary, grow arrogant, assuming that their natural talent alone will allow them to outshine their peers with minimal effort. Both perspectives are misguided. While individuals may differ in their natural aptitudes, one thing remains certain: Effort is within our control. Even if someone believes they are less naturally gifted, diligent effort can lead to remarkable progress. In fact, those who rely solely on talent but lack perseverance often find themselves surpassed by their more hardworking peers.

Knowledge is accumulated over time. Continuous learning and consistent effort inevitably lead to significant intellectual growth. I have always believed that genius is nothing more than the full realization of effort. Only through relentless study can one sharpen the mind, and only through persistent hard work can true talent emerge. I will close with an age-old saying: "Diligence overcomes inadequacy, and practice makes perfect."

Learning and Understanding*

Some young professionals engaged in scientific and technical research have asked me to share my experiences in academic study and scientific research. In truth, my understanding is also limited in scope. Here, I will simply offer my personal views on academic study attitudes and methods.

■ "From Thin to Thick" and "From Thick to Thin"

Science is an honest pursuit. To conduct scientific research, one must adopt an honest and truth-seeking attitude, free from any pretense or exaggeration. If you do not know something, admit it. If you do not understand something, acknowledge it—and then go deeper to find out why you do not understand it. Where exactly is the gap? Conversely, when you do understand something, ask yourself—what exactly do you understand? A genuine attitude toward learning begins with building a solid foundation.

Science is a rigorous discipline with strong continuity and systematic structure. If the foundational concepts are not well grasped, advancing to more complex topics becomes difficult. Many problems encountered in research often stem from weak foundational knowledge. The speed at which one progresses in scientific research is frequently tied to the strength of their foundation. Much has already been written about the importance of a strong foundation, so I will not dwell on that. Instead, I want to address a more pressing question: What should you do if you discover that your foundation is weak?

Ideally, one should establish a strong foundation from the start. But if your foundation is weak, does that mean you will never advance? Does it mean you are

* First published in issue 12 of *China Youth* in 1962.

unfit for high-level research? Should you lose confidence? Of course not. However, you must also avoid the temptation to simply "get by" without truly addressing the gaps in your knowledge. The key is to never let challenges slip by unnoticed. When we encounter something we do not understand, there are two attitudes. The first is to dismiss it—"Well, I don't get it anyway"—and move on, either out of carelessness or loss of confidence. The second is to make a genuine effort to fill in the gaps. There are two main ways to do this: One is to go back to the basics and relearn from the ground up; the other, more commonly used, is to quickly familiarize yourself with the parts you currently need, progressively filling in the missing knowledge along the way (gradually building a more complete understanding). Whatever is lacking should be reinforced through focused practice, and consistent effort should be maintained.

In this regard, the scientific community can learn from the communities of theatrical drama and Peking Opera. A veteran Peking Opera performer once said: "If I don't practice for a day, only I can tell. If I don't practice for three days, my peers can tell. If I don't practice for a month, the audience can tell." This applies just as much to scientific research. Science, like the performing arts, requires continuous accumulation and practice. Neglecting it for long periods leads to setbacks. However, if the gaps in your foundation are too significant, it is best to honestly start over rather than reaching beyond your current level—otherwise, you may find yourself in a position where you are neither proficient at the basics nor truly excelling at advanced topics.

Some may ask, "But when will this foundation-building ever end? If we keep working on fundamentals, when will it be enough?" In my view, truly solid learning requires two essential processes:

1. from thin to thick: the process of learning and reception
2. from thick to thin: the process of digestion and refinement

For example, when we read a thick book and add our own annotations, the book seems to grow even thicker, and our knowledge expands—this is the process of going "from thin to thick." However, this stage is primarily about receiving and memorizing information. Learning does not stop here, nor does true understanding. To truly grasp and master what we have learned, we must go through the process of "from thick to thin"—digesting, internalizing, and distilling the key ideas.

We often experience this: When you deeply study a book or a stack of materials, fully understanding its essence and key points, you will feel as if the book or materials have become "thinner." It may seem like you have retained less, but in reality, the knowledge has been refined and condensed into its most essential form. This is not merely accumulating information but achieving a qualitative improvement in understanding. Only through this process of digestion and refinement can we solidify our foundation. With such a strong base, further practice becomes more than just routine exercises, and reading new books is no longer about cramming information but about integrating new concepts and methods into an already well-established framework.

By following the "from thin to thick" and "from thick to thin" process, we achieve a deep, comprehensive understanding—an understanding that has been fully absorbed. Only then can we say that our foundation is truly solid. With such a foundation, future learning becomes significantly faster. This process embodies the principle of gradual progression in both education and scientific research.

Some might wonder: Does this methodical, step-by-step approach contradict ambition and the pursuit of excellence? Does it mean we should only focus on fundamentals and never strive for advanced breakthroughs? On the contrary, establishing a strong foundation is precisely the path to achieving ambitious goals. It is the only way to reach the cutting edge of science and climb the highest peaks of knowledge. Without a solid foundation, any apparent "progress" is merely superficial. True ambition must be built upon a firm base—otherwise, it is nothing but empty aspiration. Achieving great heights requires careful, systematic effort. One must climb step by step, ensuring that no step is skipped.

■ Independent Thinking and the Inheritance of Innovation

Science is not static or confined to a single standard—it is a continuous process of innovation and change. Conducting scientific research requires the ability to think independently and work autonomously. Many individuals, upon entering the field, will inevitably encounter new problems—problems that are not covered in textbooks and were never explained by teachers. In such situations, what should one do? Should we disregard these challenges simply because we haven't studied them before? Should we ask senior scientists, and if they don't have an answer, just give up? Should we search through scientific literature, and if we can't find anything relevant, abandon the problem? When neither asking nor searching provides a solution, that is precisely when independent thinking becomes crucial in finding answers.

I believe the ability to think independently should ideally be cultivated as early as possible—even during junior high school, if circumstances permit. As a vast nation engaged in the construction of a brand-new socialist society, we are bound to encounter numerous challenges that have no precedent in textbooks or past experiences. Take, for example, the Sanmenxia Dam on the Yellow River and the future Three Gorges Dam project—these are undertakings our senior scientists have never tackled before. Our mines often differ from those found abroad, making it impossible to simply replicate foreign methods. Thus, we must rely on our own research and innovation to forge a unique path forward.

Fostering independent thinking and the ability to work autonomously does not mean rejecting the achievements of our predecessors. On the contrary, it must be built upon a broad and deep understanding of their accomplishments. I greatly admire the words of Zu Chongzhi, a renowned scientist from the Five Dynasties period, who summarized his approach to learning in two key actions: search and refine from present and past. "Search" means extensively gathering knowledge from past and present sources, studying and absorbing the wisdom of predecessors. "Refine" means more than just collecting information—it requires comparison, critical analysis, digestion, and synthesis. Zu Chongzhi read widely, engaged in comparative study, and extracted key insights, ultimately developing his own theories. His work on calculating pi was not an isolated effort but rather an extension of Liu Hui's contributions, further refining and advancing the concept. Similarly, his Daming Calendar was built upon He Chengtian's Yuanjia Calendar. Many scientific and technological innovations arise from a combination of inheriting past knowledge and applying independent thought.

Independent thinking and autonomous work do not mean rejecting guidance and assistance from teachers, but they do mean not becoming overly reliant on them, either. While it is beneficial to have a mentor who can accelerate one's progress, in the course of a lifetime, direct mentorship is temporary, while self-reliance is a constant necessity. There will not always be a book that contains the exact answer one needs, and even when resources are available, knowledge often requires adaptation, creative application, or even the development of entirely new methodologies. Even under a teacher's guidance, success ultimately depends on personal effort and inquiry. Knowledge that is gained through personal struggle, contemplation, and perseverance is the most deeply ingrained and enduring. When faced with difficulties, those who have trained themselves in independent thinking will have both the confidence and capability to overcome them.

Whether fear arises in scientific research often depends on one's ability to rely on personal effort and withstand challenges. With this ability, one can face any challenge without fear. There are two types of fearlessness: The first is reckless ignorance—being unafraid simply because one doesn't understand the gravity of the problem, much like a newborn calf that does not yet fear the tiger. This kind of fearlessness is unstable and will quickly crumble when true difficulties arise. The second type, however, is cultivated through dedicated effort, persistence, and the ability to endure hardship. It is steadfast and commendable.

Young people must learn to think and work independently—to rely on their own efforts to forge a path. It may be convenient to always lean on mentors and senior scientists for guidance, but doing so exclusively can be a disadvantage. Without experiencing the rigor of independent problem-solving, knowledge remains shallow and untested, making future challenges even more daunting. That being said, this is not to downplay the invaluable role of senior scientists—their mentorship is of great benefit to the young. However, young scientists should not rely solely on their guidance; instead, they must focus on developing their own ability to think and work independently.

If young comrades have the opportunity to work alongside senior scientists, they should humbly learn from them. What should they learn? Senior scientists possess extensive knowledge and valuable experience in success, which are worth studying. However, even more important is learning from their failures—how they faced difficulties, endured setbacks, and ultimately overcame challenges. These experiences are the most precious lessons of all. One must not assume that scientific research is always smooth sailing, leading to success at every turn. Throughout history, failures in scientific research far outnumber successes. Every invention is built upon numerous failed attempts. While scientific achievements are published in newspapers, journals, and books, these reports mainly focus on successes. However, behind every success lies countless failed experiments and painstaking efforts. If we can learn well from these failures, we will become fearless. Otherwise, we may either develop an irrational fear of failure or assume that success is easily achieved.

For instance, imagine a junior high school student asking their math teacher a difficult question. The next day, the teacher writes the solution on the blackboard. While the answer is provided, something crucial is missing—the thought process behind solving the problem, much like serving a finished meal without showing the cooking. The teacher may have spent the entire night contemplating the solution, researching books, consulting references, or even staying up without sleep. If the student only sees the final answer without understanding the effort behind it, they might mistakenly believe that the teacher is simply a genius. Similarly, if we only admire the achievements of senior scientists without understanding the trials they endured, we may wrongly think that they are born with extraordinary talent while we are not.

Thus, we must learn not only from their successes but also from their failures before success. Only then can we grasp the full scope of scientific research; otherwise, we will have learned only half of it—perhaps even less. In scientific research, success is rare, while failure is common. By gaining a complete perspective, we will not be discouraged when facing difficulties.

■ Knowledge, Learning, and Experience

Human understanding follows a process that moves from sensory perception to rational comprehension. Learning and scientific research also involve a progression from "knowing" to "understanding." The concepts of "knowledge," "learning," and "experience" reflect this dual aspect of cognition and its two stages. "Knowledge" means first knowing, then understanding; "learning" means first studying, then comprehending; and "experience" means first seeing, then grasping. Simply knowing, studying, or seeing is not enough—one must refine and elevate this knowledge to the level of deeper understanding. To understand the essence of things, achieve flexible application, and make knowledge truly our own, we must progress from knowing to understanding, from learning to understanding, and from seeing to understanding—continuously improving along the way.

Confucius once said, “Learning without thought is labor lost; thought without learning is perilous.” This means that merely learning without thoughtful reflection leads to no real gain, while thinking without learning—without building upon the achievements of predecessors—is also dangerous. Both learning and thinking are indispensable. We must not only value learning but also elevate what we learn into deeper understanding.

If someone is clearly ignorant yet pretends to be insightful, or possesses only a small amount of knowledge but boasts of great wisdom, they are merely deceiving themselves. Knowledge, learning, and experience form the foundation of understanding, but true comprehension represents a higher stage. The process of advancing from knowledge to deep understanding is what Chairman Mao described as “discarding the dross and selecting the essence, eliminating the false and retaining the true, moving from the particular to the general, and penetrating from the superficial to the profound.” Without this process, one cannot achieve true comprehension.

Generally, the breadth of one’s knowledge is measured by its scope—terms like broad or vast indicate extensive knowledge. Meanwhile, the depth of one’s understanding is measured by its profundity—words like deep and refined signify true mastery. Therefore, we must strive for both breadth and depth in our knowledge. “Broad and profound” is the best compliment for true knowledge. Only when a person’s knowledge, learning, and experiences are both extensive and deeply understood can they truly be considered knowledgeable, learned, and experienced.

In ancient times, people used to praise individuals with phrases like “reading ten lines at a glance” and “never forgetting what they read,” considering them highly knowledgeable. However, in essence, such remarks merely highlight a person’s ability to learn quickly and retain information well. Without additional commendations, such a person is, at best, nothing more than a walking library or a living dictionary. If their level of insight is not high, they remain far behind those who can “infer three things from hearing one” or even “infer ten things from hearing one.” The difference is clear—one possesses reasoning ability, while the other does not.

Those who can reason have the potential to progress from mere knowledge to true understanding and may even go on to invent and create. Meanwhile, those who lack reasoning skills can only drift in the vast ocean of knowledge, submerged and unable to rise above it—let alone attain a broad and far-reaching vision. Even today, we often hear statements such as: “This student is outstanding; they completed their third-year university courses within the first two years.” Or, “This teacher is excellent; they covered a year and a half's worth of material in just one year, and all the students understood it." While these statements recognize a student's diligence or a teacher's efficiency, they fall short in assessing true excellence—because they only address the aspect of "knowing" while neglecting "understanding."

In reality, carefully reading through a few books is merely a starting point. The real goal of education is not just to absorb information in the books but to fully digest and internalize it. This is even more crucial in scientific research. Those who possess knowledge but lack true understanding cannot produce high-quality work. Being familiar with global literature in a particular field does not instantly make one a "master of knowledge" in that area—far from it! It is merely the beginning of research. Likewise, simply being able to report the latest global scientific achievements at a conference does not mean one has reached the forefront of the field. No, that is also just a starting point. The real beginning of scientific research is when one develops the ability to analyze and interpret these findings critically. The former is still a long way from producing high-level work; there is a qualitative leap still needed.

In our work, it is good to pursue knowledge and seek out new information, but we must not be satisfied with that alone. Some comrades have been working for years, and they can no longer judge themselves merely by how much they "know." Instead, they must rigorously examine whether they have elevated their knowledge and experiences to the level of true understanding. For newcomers in a field, it is not enough to simply read books and review documents—they must also be guided to comprehend, analyze, and extract the essential points from these materials. Helping them grasp the importance of progressing from "knowing" to "understanding" is critical.

The journey from knowledge, learning, and experience to true understanding is not a one-time event but a continuous process of refinement and deepening. What we believe we understand today might later reveal itself as incomplete, and we may realize that we have overlooked something more fundamental and essential. Meanwhile, as we continuously expand our knowledge, learning, and experiences, our desire for deeper understanding will drive us to greater intellectual heights. The growth of our understanding, in turn, enhances our ability to absorb new knowledge, learning, and experience, thus creating a reinforcing cycle—just like the surging tides of the Qiantang River, where each wave pushes the next, with the later waves rising even higher than the ones before.

Everything I have expressed here comes from my own reflections. I often feel troubled by the limitations of my knowledge and the inadequacy of my understanding in my work. That is why I share these thoughts with young comrades as a point of reference. I do not claim to offer any special insights, nor do I consider my views to be fully developed.

Finite and Infinite, Discrete and Continuous*,**: In Commemoration of the Fifth Anniversary of the University of Science and Technology of China

These are some reflections from our experience teaching basic mathematics to lower-year students. It seems that we have noticed certain issues, but due to the limitations of the author's knowledge, the understanding of mathematics remains partial, and without any philosophical training, I cannot derive rational conclusions from various empirical observations. Therefore, I am writing this article as a source of material, hoping that, like grains of sand forming a tower or strands of fur making a coat, it may serve as a reference for those seeking valuable insights.

Mathematics consists of two major types of problems: those of a discrete nature and those of a continuous nature. In our lifelong learning journey, we begin with counting—one, two, three, four, five … This is entirely discrete. Arithmetic and algebra are disciplines that deal with discrete problems. The mathematics taught

* This article was written in collaboration with Comrade Wang Yuan. We would like to express our gratitude to Deputy Director Pei Lisheng of the Chinese Academy of Sciences for his encouragement. He suggested that our teaching experiences should not be confined solely to mathematical works [1] or textbooks [2], but that we should also share insights that might be relevant to other related disciplines, fostering mutual exchange. It is with this intention that we have written this rather miscellaneous article, and we sincerely welcome feedback from colleagues both within and beyond our field.

** First published in the January 1963 issue of *Science Bulletin*.

throughout secondary school does not significantly emphasize "continuity" and "infinity." Although points on a straight line exhibit continuity, the explicit use of "continuity" and "infinity" truly begins with calculus. When describing instantaneous velocity or the rate of change at a given moment, we heavily rely on the concept of "continuity." This marks the boundary between elementary mathematics and advanced mathematics.

However, if one were to conclude—based on the terms "elementary" and "advanced" or the sequence in which we learn them—that "continuous mathematics" is superior to "discrete mathematics" or more capable of solving problems, that would not be entirely accurate. The purpose of this article is to highlight the importance of discreteness. It must be emphasized, however, that we are not suggesting that continuity is of lesser importance, but rather that both must be properly integrated. Everything should be determined based on practical needs—one must not overemphasize one aspect while neglecting the other. Nevertheless, there is one recommendation that may be useful for beginners: Before delving into continuous mathematics, one should first establish a solid foundation in the corresponding discrete mathematics. This is because the vast majority of continuous results are often based on discrete results or represent the limits of discrete problems. However, this does not mean that we should avoid spending more time and effort studying continuous mathematics.

Let's first consider objective reality: If something is inherently discrete, there is no need to artificially introduce continuity (though continuity should not be dismissed entirely—after all, in cases where discrete elements become too numerous to handle, a continuous approach may naturally become necessary, as in the case of flowing sand). In capitalist countries, some economists use differential equations to model economic problems. Since we have no expertise in economics, we cannot critique this approach. However, one thing is certain: The data they rely on is discrete—or fundamentally impossible to be continuous. For example, agricultural production cannot be measured in terms of instantaneous output. It follows a seasonal pattern, making even monthly breakdowns impractical, let alone finer divisions.

Is it appropriate to use continuous methods to address discrete problems? Some defend this approach by arguing that differential equations provide qualitative theories that make it easier to discern development trends. But is that really the case? In fact, trends can just as easily be identified using difference equations or matrix exponentiation, and these approaches are often simpler and more intuitive. However, it has not been included in university courses, or if it is in principle, it has not been emphasized as extensively as differential equations. In the third section, we will further discuss cases where continuity is impossible to impose, showing that forcefully applying advanced mathematics in such situations is unnecessary. True mastery lies in making complex ideas accessible, while making simple ideas convoluted is merely a waste of effort.

We hold an immature view that learning some matrix knowledge and difference equations before studying differential equations can provide methods for handling "discrete" problems, and by taking their limits, one can often derive the results of differential equations.

The above discussion emphasizes the argument that discrete problems should be handled using discrete methods. Now, let us further elaborate on the discrete processing methods in continuous problems.

The first issue is data acquisition. Is it possible to obtain infinitely precise data? No. Even if the data is accurate to ten or a hundred decimal places, the collected data set remains discrete rather than continuous (and in some cases, excessive precision is completely unnecessary). Moreover, the number of times data is collected is necessarily finite, making it inherently discrete.

Next, consider computational tools. Modern digital electronic computers are fundamentally discrete. Their characteristic is that they perform a finite number of operations on finite-bit data to obtain a finite number of finite-bit solutions. Everything is finite, and thus remains discrete.

Finally, the results that can be obtained (or are required in practice) are also discrete. This forms a process from discrete to discrete. Mathematicians usually approach this by using interpolation or regression methods to derive functions from discrete data, then obtaining differential equations. Since these differential equations often cannot be solved directly, they are discretized into algebraic equations (discrete) using difference methods, and then solved for discrete answers. Throughout this process, interpolation introduces errors, difference methods introduce additional errors, and there are further computational errors when solving algebraic equations.

This leads to the question: Can we transition directly from discrete to discrete? This would avoid errors introduced by function approximation and the discretization of differential equations. If feasible, the method would become more elementary while potentially yielding even more precise results! Given our limited expertise, we refrain from making extensive assertions, but we subjectively believe this is a direction worth exploring. However, let us reiterate: While emphasizing discrete methods, we must not overlook continuous methods. Analytic number theory, for instance, is a branch that achieves significant results by applying continuous methods to discrete problems. Considerations of continuity often reveal insights that are not easily observed through discrete analysis alone.

Below, we present several examples derived from teaching foundational courses. The selection criterion is naturally aligned with foundational-level topics or slightly advanced ones, particularly those that might be of interest to researchers in other scientific disciplines. The sections are not necessarily closely related. For instance, those familiar with Fourier series might find section 4 particularly relevant.

To reiterate, this article is meant to serve as a catalyst for discussion. By providing concrete empirical examples, we hope to create conditions for future educational reforms and theoretical developments. We sincerely seek guidance and welcome any corrections.

■ The Object Is Continuous but We Can Only Obtain Finite Data—Calculating Volume and Area

Having learned calculus, we often feel a sense of joy: Any curve length, any surface area, and any object's volume can now be handled using integration methods. This joy is well-founded and understandable. However, if we believe that this alone is enough to solve all problems, we would be mistaken. Upon deeper reflection, we realize that the methods we have learned all share a common requirement: the need for an explicit formula representing the curve or surface. But in real-world applications, do such expressions always exist?

For example, when estimating the reserves of a mineral deposit, is there an analytical formula that represents the boundary of the mineral deposit? Or when estimating the area of a slope, is there a formula $z = f(x,y)$ that represents the surface? In practical situations, such expressions do not exist. First, because it is impossible to take measurements at every single point, and second, even if many points of a mineral deposit are measured, deriving a complete expression for the surface is still infeasible. Even if an equation is somehow pieced together, in most cases, the integral cannot be computed explicitly (i.e., an antiderivative cannot be found), and only in very rare cases can it be expressed as an elementary function. Therefore, although the mineral deposit and slope are continuously distributed, we must still rely on discrete methods to estimate their volume and area approximately.

However, this does not mean that the formulas for calculating area and volume in calculus are useless. Rather, we must consider how they can be applied effectively. It turns out that the theoretical foundations provided by calculus are indeed useful—they offer concrete guidelines, help us assess the advantages and disadvantages of different methods, and allow us to improve these methods further.

Let's look at an example. When estimating the area of a slope, two methods are commonly used: **Волков** method, which originates from geography; and **Бауман** method, which originates from ore geometry. We will introduce both methods and then compare their pros and cons.

Assume that contour lines are drawn on a map at height intervals of Δh. Further assume that there is a peak and that the contour lines form closed loops around it. (Other cases can be derived from this setup.) Denote the contour lines from the highest (l_n) to the lowest as (l_{n-1}), (l_{n-2}), …, (l_0), where (l_0) is at height 0 and (l_n) is at height h. Let B_i represent the projected area between contours (l_i) and (l_{i+1}) (fig. 1).

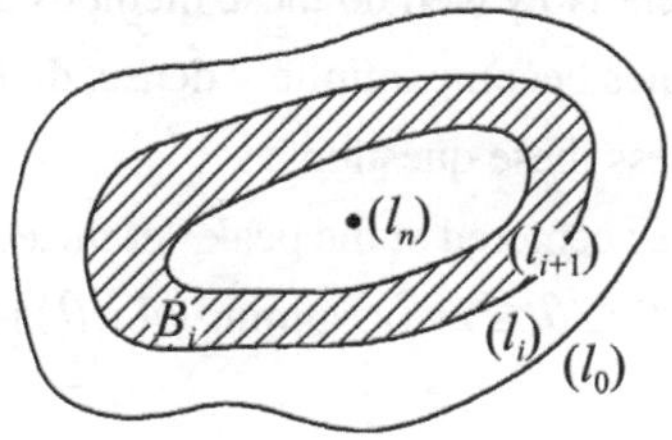

Figure 1

Бауман Method: 1. Compute

$$C_i = \frac{1}{2}(l_i + l_{i+1})\Delta h$$

which represents the area of an intermediate vertical section.[1] 2. Then, the approximate slope area is given by

$$\sum_{i=0}^{n-1}\sqrt{B_i^2+C_i^2}$$

Волков Method: 1. Compute

$$l=\sum_{i=0}^{n-1}l_i$$

(the total contour line length) and

$$B=\sum_{i=0}^{n-1}B_i$$

(the total projected area). Then, using

$$\operatorname{tg}\alpha=\frac{\Delta h\cdot l}{B}$$

find the average slope angle α. 2. Compute the approximate slope area as $B\sec\alpha=\sqrt{B^2+(\Delta h\cdot l)^2}$.

Which method is better? How well do these methods approximate the true slope area? When the contour lines become infinitely dense, do the results converge to the actual area? We now address these questions.

Using polar coordinates centered at the peak, let the equation of the contour line at height z be $\rho=\rho(z,\theta)$, $0\le\theta\le 2\pi$ (assuming $\rho(z,\theta)$ is sufficiently smooth).

Define

$$z_i=\frac{h}{n}i,\ \Delta h=\frac{h}{n}$$

Then, the area enclosed by (l_i) is

$$\frac{1}{2}\int_0^{2\pi}\rho^2(z_i,\theta)\mathrm{d}\theta$$

1 The length of the contour line (l_i) is denoted as l_i.

The length of (l_i) is

$$l_i = \int_0^{2\pi} \sqrt{\rho^2(z_i,\theta) + \left[\frac{\partial\rho(z_i,\theta)}{\partial\theta}\right]^2}\, d\theta$$

Applying the mean-value theorem, we obtain

$$B_i = -\int_0^{2\pi} \rho(z_i',\theta)\frac{\partial\rho(z_i',\theta)}{\partial z_i'}\, d\theta\Delta h$$

and

$$C_i = \int_0^{2\pi} \sqrt{\rho^2(z_i'',\theta) + \left[\frac{\partial\rho(z_i'',\theta)}{\partial\theta}\right]^2}\, d\theta\Delta h$$

where $z_i \le z', z_i'' \le z_{i+1}$.

Thus, as $\Delta h \to 0$, the sum

$$\sum_{i=0}^{n-1} \sqrt{B_i^2 + C_i^2}$$

approaches the integral

$$\text{Ба} = \int_0^h \sqrt{\left(\int_0^{2\pi} \rho\frac{\partial\rho}{\partial z} d\theta\right)^2 + \left[\int_0^{2\pi} \sqrt{\rho^2 + \left(\frac{\partial\rho}{\partial\theta}\right)^2} d\theta\right]^2}\, dz$$

This is the limiting value of the slope area estimated using Бауман method when $\Delta h \to 0$.

Similarly, the limit of

$$\sqrt{\left(\sum_{i=0}^{n-1} B_i\right)^2 + \left(\Delta h \sum_{i=0}^{n-1} l_i\right)^2}$$

is

$$\text{Bo} = \sqrt{\left(\int_0^{2\pi} d\theta \int_0^h \rho\frac{\partial\rho}{\partial z} dz\right)^2 + \left[\int_0^h dz \int_0^{2\pi} \sqrt{\rho^2 + \left(\frac{\partial\rho}{\partial\theta}\right)^2} d\theta\right]^2}$$

This is the limiting value of the slope area estimated using Волков method.

The known surface area S is given by

$$S=\int_0^h\int_0^{2\pi}\sqrt{\rho^2+\left(\frac{\partial\rho}{\partial\theta}\right)^2+\left(\rho\frac{\partial\rho}{\partial z}\right)^2}\,\mathrm{d}\theta\mathrm{d}z$$

Introducing the complex function

$$f(z,\theta)=-\rho\frac{\partial\rho}{\partial z}+i\sqrt{\rho^2+\left(\frac{\partial\rho}{\partial\theta}\right)^2}$$

we obtain

$$S=\int_0^h\int_0^{2\pi}|f(z,\theta)|\mathrm{d}\theta\mathrm{d}z,\ \text{Ба}=\int_0^h\left|\int_0^{2\pi}f(z,\theta)\mathrm{d}\theta\right|\mathrm{d}z,\ \text{Bo}=\left|\int_0^h\int_0^{2\pi}f(z,\theta)\mathrm{d}\theta\mathrm{d}z\right|$$

So Bo $\leq$ Ба $\leq S$.

Conclusions: 1. Бауман method is more precise than Волков method. 2. The results obtained using these methods tend to underestimate the true area.

Apart from this, it is not difficult to discuss the cases where Bo $= S$ and Ба $= S$. We can also provide an estimate of the errors introduced by these methods, identify the causes of these errors, and suggest ways to avoid them. For more details, please refer to [1] and [2].

Note 1: The integrals used in this section can be avoided.

■ Impossible to Continuize: Non-Negative Matrices

Quantities such as production, energy, and probability cannot be negative. In various problems related to cosmic ray showers, operations research, and probability theory, non-negative element matrices frequently appear. These matrices express changes in the quantities of certain physical states over a given time period.

To be more specific, suppose there are three types of substances—A, B, and C—each initially present in amounts a, b, and c. After a certain time t, the amount of type A transforms into ap_{11}, ap_{12}, and ap_{13} units of substances A, B, and C, respectively. Similarly, type B transforms into bp_{21}, bp_{22}, and bp_{23} units of substances A, B, and C, and type C into cp_{31}, cp_{32}, and cp_{33} units of substances A, B, and C. Thus, after time t, the amounts of substances A, B, and C become as $ap_{11}+bp_{21}+cp_{31}$, $ap_{12}+bp_{22}+cp_{32}$, $ap_{13}+bp_{23}+cp_{33}$.

Since the amount of a substance cannot become negative, we have $p_{ij} \geq 0$. The matrix

$$P = \begin{pmatrix} p_{11} & p_{12} & p_{13} \\ p_{21} & p_{22} & p_{23} \\ p_{31} & p_{32} & p_{33} \end{pmatrix}$$

is called the transition matrix. If the initial amounts of the substances are represented as a vector $\boldsymbol{v} = (a, b, c)$, then after time t, their quantities are given by $\boldsymbol{vP}$.

If this transformation process continues in the same manner, then after $2t$, we obtain $\boldsymbol{vP}^2$.

After nt time units, we get $\boldsymbol{vP}^n$.

By increasing n, we can observe the trend of development.

To better understand the implications, let's consider a single-variable case. Suppose an initial quantity c transforms into cq after one unit of time, and after n units of time, it becomes cq^n. We might assume that after half a unit of time, the quantity is $cq^{1/2}$ (this assumption is where the issue arises). More generally, at time t, the quantity can be modeled as $f(t) = cq^t$. Its differential equation is

$$\frac{\mathrm{d}f}{\mathrm{d}t} = (\log q)f, \ f(0) = c$$

Thus, $f(t) = cq^t$ is the unique solution of this differential equation.

For a single phenomenon, even though there is a theoretical inconsistency—whether at $t = 1/2$ the amount is indeed $q^{1/2}$ times the original—the issue does not pose practical difficulties. This is because, for a positive number, any root can be taken, i.e.,

$$\lim_{\varepsilon \to 0} \frac{q^{\varepsilon} - 1}{\varepsilon} = \log q$$

exists.

This principle suggests that the rate of quantity increase is proportional to time. Therefore, even if $q^{1/2}$ is not rigorously defined, assuming its existence in an idealized manner does not lead to contradictions.

If someone wishes to extend this principle to multiple phenomena, it inevitably requires finding the square root of matrix $\boldsymbol{P}$, or even arbitrary roots of $\boldsymbol{P}$. The key question is: Does there exist a non-negative matrix whose square equals $\boldsymbol{P}$? If no such matrix exists, then using differential methods becomes impossible.

For example, there is no non-negative matrix whose square equals

$$\begin{pmatrix} 0 & 1 \\ 1 & 0 \end{pmatrix}$$

The reason is simple: If

$$\begin{pmatrix} a & b \\ c & d \end{pmatrix}^2 = \begin{pmatrix} 0 & 1 \\ 1 & 0 \end{pmatrix}$$

then expanding the matrix multiplication gives the equation: $a^2 + bc = 0$. Since a, b, and c are non-negative, this implies that $a = 0$ and b or $c = 0$, which is impossible.

In other words, if the matrix $\boldsymbol{P}$ is not "infinitely divisible," meaning that there does not exist a real matrix $\boldsymbol{Q}$ such that $\boldsymbol{P} = \mathrm{e}^{\boldsymbol{Q}}$, then it is impossible to handle it using differential methods. In the study of oscillations in linear elastic systems, the eigenvalues of the associated matrix are all positive and distinct, which guarantees the existence of $\boldsymbol{Q}$. However, in economic models, where oscillatory progress and spiral growth occur, the corresponding matrices may have negative or complex roots. The presence of negative roots means that the matrix is not infinitely divisible, making differential equation modeling impractical. Therefore, blindly applying differential equations to economic phenomena can be counterproductive.

For physical and probabilistic models, when using differential equations, we must first assess whether they can be used and whether they should be used. If infinite divisibility cannot be proven, difference equations provide a safer alternative. Even when infinite divisibility is established, difference equations may still be the simpler choice. Differential equations are not always necessary.

Here is a bit of an aside: Does there exist a fundamental "unit" of time required to complete a transformation? That is, if the time interval is shorter than this unit, the transformation cannot be completed. In such cases, the difference equation method may better represent objective phenomena than the differential equation method. In these scenarios, time becomes discrete. But how long is this fundamental unit? What happens if multiple processes with different fundamental units are mixed together? Mathematically, this leads to questions of commensurability and incommensurability. As a result, phenomena similar to Diophantine approximation in number theory emerge—but in a much more complex form.

Note 1: some properties of non-negative matrices.

Theorem 1: If

$$\sum_{i=1}^{n} a_{ij} \leq q, a_{ij} \geq 0, \ j = 1,2,\ldots,n \tag{1}$$

Then, the absolute values of the eigenvalues of the matrix $\boldsymbol{A} = (a_{ij})$ are all $\leq q$.

The proof of this theorem is quite simple. By the definition of the eigenvalue λ, there exist nonzero $x_1,\ldots,x_n$ such that

$$\sum_{j=1}^{n} a_{ij} x_j = \lambda x_i$$

Thus,

$$|\lambda||x_i| \leq \sum_{j=1}^{n} a_{ij} |x_j|$$

So,

$$|\lambda| \sum_{i=1}^{n} |x_i| \leq \sum_{j=1}^{n} \left(\sum_{i=1}^{n} a_{ij} \right) |x_j| \leq q \sum_{j=1}^{n} |x_j|$$

proving the claim.

This property will be useful later, so a proof has been provided. In essence, several properties and characteristics of nonnegative matrices seem to have significant implications in economics (or other disciplines where they are applicable). For example, a nonnegative matrix has a largest positive eigenvalue, which can serve as an indicator of the growth rate of an economic system. Its corresponding eigenvector with nonnegative elements appears to reflect the correct equivalence relationships between different products or between products and labor. If complex eigenvalues exist, it suggests the possibility of spiral growth or wave-like progress in certain sectors.

Moreover, this analysis can offer insights into which coefficients (e.g., the coal consumption per ton of steel) should be improved to maximize the growth rate of the economic system. Identifying the key factors for improvement is crucial. Of course, such recommendations should only be considered as references—ultimately, human decisions play the most important role.

■ Overcalculating Leads to Greater Errors—Practical Harmonic Analysis

In many applications, we frequently need to expand a function $f(x)$ into a Fourier series, given by

$$f(x) \sim \frac{a_0}{2} + \sum_{m=1}^{+\infty} (a_m \cos mx + b_m \sin mx) \tag{1}$$

where

$$a_m = \frac{1}{\pi}\int_0^{2\pi} f(x)\cos mx\mathrm{d}x,\ b_m = \frac{1}{\pi}\int_0^{2\pi} f(x)\sin mx\mathrm{d}x \tag{2}$$

Alternatively, we can use the equivalent complex form of the Fourier series:

$$f(x) \sim \sum_{m=-\infty}^{+\infty} C_m \mathrm{e}^{imx},\ C_m = \frac{1}{2\pi}\int_0^{2\pi} f(x)\mathrm{e}^{-imx}\mathrm{d}x \tag{3}$$

If $f(x)$ is obtained from experiments, sometimes only a finite number of data points are measured. Based on these finite data points, how can we obtain an approximate Fourier series? Additionally, even if $f(x)$ has an analytic expression, the antiderivatives of the integrals in (2) and (3) may not be obtainable, necessitating numerical integration. In both cases, a common approach is used.

Assuming that the function values are given at $n(= 2n' + 1)$ points in the interval $[0,2\pi]$:

$$y_l = f\left(\frac{2\pi l}{n}\right),\ 0 \le l \le n-1$$

we approximate a_m and as b_m

$$a_m \sim a'_m = \frac{2}{n}\sum_{l=0}^{n-1} y_l \cos\left(\frac{2\pi lm}{n}\right)$$

$$b_m \sim b'_m = \frac{2}{n}\sum_{l=0}^{n-1} y_l \sin\left(\frac{2\pi lm}{n}\right)$$

It might seem intuitive that using fewer data points while calculating more terms of a'_m and b'_m using modern computational tools would result in a more precise approximation:

$$\frac{a_0}{2}+\sum_{m=1}^{N}(a'_m\cos mx+b'_m\sin mx) \tag{4}$$

However, this is incorrect. If only n data points are provided—meaning that a_m and b_m are approximated using only n function values—overcalculating does not increase accuracy but instead increases errors, even leading to absurd conclusions. The reason is $a'_m=a'_{n+m}, b'_m=b'_{n+m}\,(m=1,2,\ldots)$.

Thus, the series

$$\frac{a'_0}{2}+\sum_{m=1}^{+\infty}(a'_m\cos mx+b'_m\sin mx) \tag{5}$$

diverges. As a result, continuing the summation will lead to a result that significantly deviates from the original function $f(x)$, particularly if $f(x)$ is a sufficiently smooth function (e.g., continuous and differentiable). We can prove that the best approach is to sum up to n terms. Summing beyond that wastes effort and increases errors, while summing fewer terms fails to fully utilize the available data.

A simple discrete exponential sum approach is used to address this issue. The method is discrete and easy to compute, starting from the complex form of the Fourier series. Assume that $n(=2n'+1)$ data points of a function $f(x)$ are given in the interval $[-\pi,\pi]$:

$$y_l=f\left(\frac{2\pi l}{n}\right),\ l=0,\pm 1,\ldots,\pm n' \tag{6}$$

Using the formula

$$\frac{1}{n}\sum_{l=-n'}^{n'}e^{2\pi ilm/n}=\begin{cases}0, \text{if } n\nmid m\\ 1, \text{if } n\mid m\end{cases} \tag{7}$$

we derive C'_m from

$$y_l=\sum_{m=-n'}^{n'}C'_m e^{2\pi ilm/n},\ |l|\le n' \tag{8}$$

To determine C'_m, we multiply both sides of (8) by $e^{-2\pi i lq/n}$ and sum over l. Using (7), we obtain

$$\sum_{l=-n'}^{n'} y_l e^{-2\pi i lq/n} = \sum_{m=-n'}^{n'} C'_m \sum_{l=-n'}^{n'} e^{2\pi i(m-q)l/n} = nC'_q \tag{9}$$

which suggests the approximation

$$\begin{aligned} S_n(x) &= \sum_{m=-n'}^{n'} C'_m e^{imx} \\ C'_m &= \frac{1}{n}\sum_{l=-n'}^{n'} y_l e^{-2\pi i lm/n} \end{aligned} \tag{10}$$

Now, we estimate the error between $S_n(x)$ and $f(x)$.

Theorem 1: Assume that $f(x)$ has $r(\geq 2)$-th order continuous derivatives in $[-\pi,\pi]$ and is 2π-periodic. If $|f^{(r)}(x)| < C$, then

$$|f(x)-S_n(x)| < \frac{4C}{(r-1)n'^{r-1}} \tag{11}$$

Proof: Given that

$$\begin{aligned} f(x) &= \sum_{m=-\infty}^{+\infty} C_m e^{imx} \\ C_m &= \frac{1}{2\pi}\int_{-\pi}^{\pi} f(x)e^{-imx}\,dx \end{aligned} \tag{12}$$

Integrating by parts r times gives

$$C_m = \frac{1}{2\pi(im)^r}\int_{-\pi}^{\pi} f^{(r)}(x)e^{-imx}\,dx$$

which immediately yields

$$|C_m| \leq \frac{C}{|m|^r}$$

Thus,

$$\left| f(x) - \sum_{m=-n'}^{n'} C_m e^{imx} \right| \leq 2 \sum_{m=n'+1}^{+\infty} \frac{C}{|m|^r}$$
$$\leq 2C \int_{n'}^{\infty} \frac{dx}{x^r} = \frac{2C}{(r-1)n'^{r-1}} \tag{13}$$

When $|m| \leq n'$, we have

$$C_m - C'_m = C_m - \frac{1}{n} \sum_{l=-n'}^{n'} y_l e^{-2\pi i l m/n}$$
$$= C_m - \frac{1}{n} \sum_{l=-n'}^{n'} e^{-2\pi i l m/n} \sum_{q=-\infty}^{+\infty} C_q e^{2\pi i q l/n}$$
$$= C_m - \frac{1}{n} \sum_{q=-\infty}^{+\infty} C_q \sum_{l=-n'}^{n'} e^{2\pi i (q-m)l/n}$$
$$= C_m - \sum_{\substack{q=-\infty \\ q \equiv m (\bmod n)}}^{+\infty} C_q$$

Therefore,

$$|C_m - C'_m| \leq \sum_{t=-\infty}^{+\infty}{}' |C_{m+nt}|$$

where $\sum'$ denotes the summation excluding the term $t = 0$. Hence,

$$\left| \sum_{m=-n'}^{n'} (C_m - C'_m) e^{imx} \right| \leq \sum_{m=-n'}^{n'} \sum_{t=-\infty}^{+\infty}{}' |C_{m+nt}|$$
$$\leq \sum_{m=-n'}^{n'} \sum_{t=-\infty}^{+\infty} \frac{C}{|m+nt|^r}$$
$$\leq 2C \sum_{t=n'+1}^{+\infty} \frac{1}{l^r} \tag{14}$$
$$\leq \frac{2C}{(r-1)n'^{r-1}}$$

(Any integer l can be uniquely written as $nt + m(|m| \leq n')$ but with $t \neq 0$, which expresses the removal of all integers $|l| \leq n'$, thus proving the claim.)

Therefore, from (12), (13), and (14), we obtain

$$|f(x) - S_n(x)| < \frac{4C}{(r-1)n'^{r-1}}$$

In actual calculations, $S_n(x)$ can be expressed in a simpler form:

$$\begin{aligned}
S_n(x) &= \sum_{m=-n'}^{n'} C'_m \mathrm{e}^{imx} = \frac{1}{n}\sum_{m=-n'}^{n'}\sum_{l=-n'}^{n'} y_l \mathrm{e}^{-2i\pi lm/n}\mathrm{e}^{imx} \\
&= \frac{1}{n}\sum_{l=-n'}^{n'} y_l \sum_{m=-n'}^{n'} \mathrm{e}^{i\left(x-\frac{2\pi l}{n}\right)m} \\
&= \frac{1}{n}\sum_{l=-n'}^{n'} y_l \frac{\sin\left(n'+\frac{1}{2}\right)(x-2\pi l/n)}{\sin\frac{1}{2}(x-2\pi l/n)} \\
&= \frac{1}{n}\sum_{l=-n'}^{n'} y_l \frac{\sin\left(\frac{1}{2}nx-\pi l\right)}{\sin\frac{1}{2}(x-2\pi l/n)} \\
&= \frac{\sin\frac{1}{2}nx}{n}\sum_{l=-n'}^{n'} \frac{(-1)^l y_l}{\sin\frac{1}{2}(x-2\pi l/n)}
\end{aligned} \qquad (15)$$

Note 1: If the sample points $0 \le x_1 < \ldots < x_n < 2\pi$ are not uniformly spaced, then by solving the system of equations:

$$\begin{cases}
\dfrac{a'_0}{2} + \sum\limits_{m=1}^{n'}[a'_m \cos mx_i + b'_m \sin mx_i] = y_i, (1 \le i \le n) \\
\dfrac{a'_0}{2} + \sum\limits_{m=1}^{n'}[a'_m \cos mx + b'_m \sin mx] = y(x)
\end{cases}$$

we can eliminate a'_0, a'_m, b'_m to derive the relationship between y and $y_1, \ldots, y_n$.

■ Finite Difference Method—Common Bridge Between Continuous and Discrete Domains

In solving differential equations, the finite difference method is frequently and widely used. Simply put, this method transforms the problem of solving differential equations into solving algebraic equations (i.e., difference equations). To illustrate this, we briefly introduce how this method is applied to the Dirichlet problem for

Laplace equation. While solving the difference equations, we will discuss both algebraic methods and the Monte Carlo method, comparing their advantages and limitations.

Problem: Let G be a bounded, simply connected region in the plane with a smooth boundary Γ. Suppose a continuous function $f(x,y)$ is given on Γ. We seek a continuous function $u(x,y)$ that satisfies:

1. Laplace equation in G:

$$\frac{\partial^2 u}{\partial x^2}+\frac{\partial^2 u}{\partial y^2}=0 \tag{1}$$

2. On Γ, take the given function f as the boundary values.

The steps to approximate u are as follows:

Step 1: grid generation. We construct a grid in the plane using lines parallel to the coordinate axes: $x=mh$, $y=nh$, where h is a positive number, and m, n take integer values. The domain G is covered by small squares with side length h, as shown in figure 2. The vertices of these squares are called grid points. The region covered by squares intersecting G is denoted as G^*, which is a polygonal approximation of G^*.

Let Q be a grid point on the boundary of G^*. We assume the nearest point on Γ to Q is P (if there are multiple nearest points, any of them can be chosen). We define $f(Q)=f(P)$. Thus, the function values $f(Q)$ are assigned to the integer points on the boundary Γ of G^*.

Step 2: finite difference approximation. By replacing the second-order partial derivatives

$$\frac{\partial^2 u}{\partial x^2}, \frac{\partial^2 u}{\partial y^2}$$

with

$$\frac{u(x+h,y)-2u(x,y)+u(x-h,y)}{h^2} \quad \text{and} \quad \frac{u(x,y+h)-2u(x,y)+u(x,y-h)}{h^2}$$

respectively, the Laplace equation can be rewritten as

$$u(x,y)=\frac{1}{4}[u(x+h,y)+u(x-h,y)+u(x,y+h)+u(x,y-h)] \qquad (2)$$

This equation means that at any interior grid point (x,y), the function value $u(x,y)$ is the average of its four nearest neighbors: east, west, north, and south (fig. 3).

Step 3: solving the linear system. The problem is now transformed into finding function values at interior grid points, given the function values at boundary grid points. This leads to a system of linear equations of the form (equation 2).

Two key questions arise: Does the system of equations obtained from this lead to a contradictory system? Is the solution unique? These are essential questions that must be answered. Now, let's start with a simple example, and then we can intuitively understand the general theory.

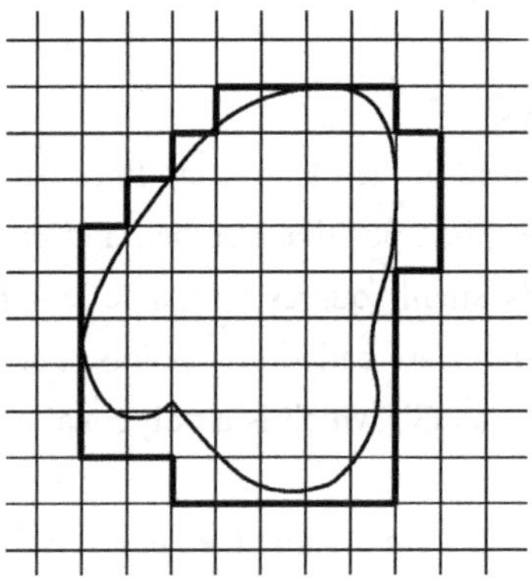

Figure 2

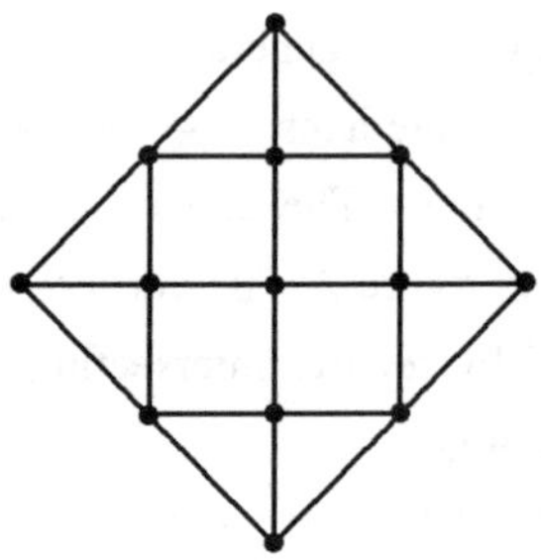

Figure 3

Let us take $h = 1$. Given the function values at eight points: $u(2,0), u(1,1), u(0,2), u(-1,1), u(-2,0), u(-1,-1), u(0,-2), u(1,-1)$, find the values of $u(0,0), u(1,0), u(0,1), u(-1,0), u(0,-1)$.[2]

Expanding equations at each interior point gives

2 It is not difficult to see that, for this example, figure 3 and the diagram  are equivalent.

$$\left.\begin{aligned}
u(0,0)&=\frac{1}{4}[u(1,0)+u(0,1)+u(-1,0)+u(0,-1)]\\
u(1,0)&=\frac{1}{4}[u(2,0)+u(1,1)+u(0,0)+u(1,-1)]\\
u(0,1)&=\frac{1}{4}[u(1,1)+u(0,2)+u(-1,1)+u(0,0)]\\
u(-1,0)&=\frac{1}{4}[u(0,0)+u(-1,1)+u(-2,0)+u(-1,-1)]\\
u(0,-1)&=\frac{1}{4}[u(1,-1)+u(0,0)+u(-1,-1)+u(0,-2)]
\end{aligned}\right\}\qquad(3)$$

Using elimination, we derive

$$\begin{aligned}
u(0,0)=&\frac{1}{12}[u(2,0)+u(0,2)+u(-2,0)+u(0,-2)]+\\
&\frac{1}{6}[u(1,1)+u(-1,1)+u(-1,-1)+u(1,-1)]
\end{aligned}\qquad(4)$$

$$\begin{aligned}
u(1,0)=&\frac{13}{48}u(2,0)+\frac{7}{24}[u(1,1)+u(1,-1)]+\\
&\frac{1}{24}[u(-1,1)+u(-1,-1)]+\\
&\frac{1}{48}[u(0,2)+u(-2,0)+u(0,-2)]
\end{aligned}\qquad(5)$$

The meaning of these coefficients $\frac{1}{6},\frac{1}{12},\frac{13}{48},\ldots$ will be explained later. First, let's perform the following algebraic manipulation: treating the function values at these thirteen points as elements of a column vector, equation (3) can be rewritten as

$$\begin{pmatrix} u(0,0) \\ u(1,0) \\ u(0,1) \\ u(-1,0) \\ u(0,-1) \\ u(1,1) \\ u(-1,1) \\ u(-1,-1) \\ u(1,-1) \\ u(2,0) \\ u(0,2) \\ u(-2,0) \\ u(0,-2) \end{pmatrix} = \left(\begin{array}{ccccc:cccccccc} 0 & \frac{1}{4} & \frac{1}{4} & \frac{1}{4} & \frac{1}{4} & 0 & 0 & 0 & 0 & 0 & 0 & 0 & 0 \\ \frac{1}{4} & 0 & 0 & 0 & 0 & \frac{1}{4} & 0 & 0 & \frac{1}{4} & \frac{1}{4} & 0 & 0 & 0 \\ \frac{1}{4} & 0 & 0 & 0 & 0 & \frac{1}{4} & \frac{1}{4} & 0 & 0 & 0 & \frac{1}{4} & 0 & 0 \\ \frac{1}{4} & 0 & 0 & 0 & 0 & 0 & \frac{1}{4} & \frac{1}{4} & 0 & 0 & 0 & \frac{1}{4} & 0 \\ \frac{1}{4} & 0 & 0 & 0 & 0 & 0 & 0 & \frac{1}{4} & \frac{1}{4} & 0 & 0 & 0 & \frac{1}{4} \\ \hdashline 0 & 0 & 0 & 0 & 0 & 1 & 0 & 0 & 0 & 0 & 0 & 0 & 0 \\ 0 & 0 & 0 & 0 & 0 & 0 & 1 & 0 & 0 & 0 & 0 & 0 & 0 \\ 0 & 0 & 0 & 0 & 0 & 0 & 0 & 1 & 0 & 0 & 0 & 0 & 0 \\ 0 & 0 & 0 & 0 & 0 & 0 & 0 & 0 & 1 & 0 & 0 & 0 & 0 \\ 0 & 0 & 0 & 0 & 0 & 0 & 0 & 0 & 0 & 1 & 0 & 0 & 0 \\ 0 & 0 & 0 & 0 & 0 & 0 & 0 & 0 & 0 & 0 & 1 & 0 & 0 \\ 0 & 0 & 0 & 0 & 0 & 0 & 0 & 0 & 0 & 0 & 0 & 1 & 0 \\ 0 & 0 & 0 & 0 & 0 & 0 & 0 & 0 & 0 & 0 & 0 & 0 & 1 \end{array}\right) \begin{pmatrix} u(0,0) \\ u(1,0) \\ u(0,1) \\ u(-1,0) \\ u(0,-1) \\ u(1,1) \\ u(-1,1) \\ u(-1,-1) \\ u(1,-1) \\ u(2,0) \\ u(0,2) \\ u(-2,0) \\ u(0,-2) \end{pmatrix} \tag{6}$$

It can be abstracted as

$$\begin{pmatrix} \boldsymbol{u} \\ \boldsymbol{v} \end{pmatrix} = \begin{pmatrix} \boldsymbol{P} & \boldsymbol{Q} \\ \boldsymbol{O} & \boldsymbol{I} \end{pmatrix} \begin{pmatrix} \boldsymbol{u} \\ \boldsymbol{v} \end{pmatrix} \tag{7}$$

Here, $\boldsymbol{v}$ is the column vector consisting of the function values at the boundary grid points, while $\boldsymbol{u}$ is the column vector consisting of the function values at the "interior grid points." This formulation applies to general problems and essentially expresses two key points: 1. The function value at an interior grid point can be represented as the average of its four neighboring values (east, south, west, and north). 2. Boundary points remain boundary points.

Since each grid point can have at most four neighboring grid points, the sum of the elements in each column of matrix $\boldsymbol{P}$ is at most 1. Now, we prove that the sum of the elements in each column of $\boldsymbol{P}^2$ is strictly less than 1. If this were not the case, then since the elements of $\boldsymbol{P}$ can only take values of either 0 or 1/4, $\boldsymbol{P}$ would necessarily contain a submatrix of the form:

$$\begin{pmatrix} \frac{1}{4} & \frac{1}{4} & \frac{1}{4} & \frac{1}{4} \\ \frac{1}{4} & \frac{1}{4} & \frac{1}{4} & \frac{1}{4} \\ \frac{1}{4} & \frac{1}{4} & \frac{1}{4} & \frac{1}{4} \\ \frac{1}{4} & \frac{1}{4} & \frac{1}{4} & \frac{1}{4} \end{pmatrix}$$

However, since no two distinct grid points can have the same four neighboring points (east, south, west, and north), this situation is impossible. Therefore, the sum of the elements in each column of $\boldsymbol{P}^2$ is strictly less than 1. Consequently, the absolute values of the eigenvalues of $\boldsymbol{P}^2$ are all less than 1, which leads to

$$\lim_{n\to\infty} \boldsymbol{P}^n = \boldsymbol{0}$$

Moreover, the series $\boldsymbol{Q}+\boldsymbol{PQ}+\boldsymbol{P}^2\boldsymbol{Q}+\ldots$ converges to $(\boldsymbol{I}-\boldsymbol{P})^{-1}\boldsymbol{Q}$. By iterating equation (7) n times, we obtain

$$\begin{pmatrix} \boldsymbol{u} \\ \boldsymbol{v} \end{pmatrix} = \begin{pmatrix} \boldsymbol{P} & \boldsymbol{Q} \\ \boldsymbol{O} & \boldsymbol{I} \end{pmatrix}^n \begin{pmatrix} \boldsymbol{u} \\ \boldsymbol{v} \end{pmatrix} = \begin{pmatrix} \boldsymbol{P}^n & \boldsymbol{Q}+\boldsymbol{PQ}+\boldsymbol{P}^2\boldsymbol{Q}+\cdots+\boldsymbol{P}^{n-1}\boldsymbol{Q} \\ \boldsymbol{O} & \boldsymbol{I} \end{pmatrix} \begin{pmatrix} \boldsymbol{u} \\ \boldsymbol{v} \end{pmatrix}$$

Taking the limit as $n \to \infty$, we get

$$\begin{pmatrix} \boldsymbol{u} \\ \boldsymbol{v} \end{pmatrix} = \begin{pmatrix} \boldsymbol{O} & \boldsymbol{Q}+\boldsymbol{PQ}+\boldsymbol{P}^2\boldsymbol{Q}+\cdots \\ \boldsymbol{O} & \boldsymbol{I} \end{pmatrix} \begin{pmatrix} \boldsymbol{u} \\ \boldsymbol{v} \end{pmatrix}$$

Thus,

$$\boldsymbol{u} = (\boldsymbol{Q}+\boldsymbol{PQ}+\boldsymbol{P}^2\boldsymbol{Q}+\cdots)\,\boldsymbol{v} = (\boldsymbol{I}+\boldsymbol{P}+\boldsymbol{P}^2+\cdots)\boldsymbol{Qv} \tag{8}$$

This provides the solution to the problem. That is, given the function values at the boundary grid points, $\boldsymbol{v}$, we can compute the function values at the interior grid points, $\boldsymbol{u}$, using equation (8). This suggests the following algorithm:

1. Algebraic method. Express $\boldsymbol{u}$ as a column vector $(u_1, \ldots, u_l)'$, and v as a column vector $(v_1, \ldots, v_k)'$. If an interior grid point[3] u_i is adjacent to a boundary grid point v_i, then set the corresponding element $\boldsymbol{Q}\,(i,j)$ to 1/4; otherwise, set it to 0. Similarly, if two interior grid points u_i and u_j are adjacent, set $\boldsymbol{P}\,(i,j)$ to 1/4; other-

3 The internal grid points and boundary grid points are denoted as $u_1, \ldots, u_l$ and $v_1, \ldots, v_k$, respectively; please do not confuse them.

wise, set it to 0. With $\boldsymbol{P}$ and $\boldsymbol{Q}$ determined, we can compute equation (8) using the following format:

	$\boldsymbol{Qv}$	
$\boldsymbol{P}$	$\boldsymbol{PQv}$	$\boldsymbol{R}_1 = \boldsymbol{Qv} + \boldsymbol{PQv}$
$\boldsymbol{P}^2$	$\boldsymbol{P}^2\boldsymbol{R}_1$	$\boldsymbol{R}_2 = \boldsymbol{R}_1 + \boldsymbol{P}^2\boldsymbol{R}_1$
$\boldsymbol{P}^4$	$\boldsymbol{P}^4\boldsymbol{R}_2$	$\boldsymbol{R}_3 = \boldsymbol{R}_2 + \boldsymbol{P}^4\boldsymbol{R}_2$
$\boldsymbol{P}^8$	$\boldsymbol{P}^8\boldsymbol{R}_3$	$\boldsymbol{R}_4 = \boldsymbol{R}_3 + \boldsymbol{P}^8\boldsymbol{R}_3$
$\vdots$	$\vdots$	$\vdots$

In our example, since

$$\boldsymbol{P}^3 = \frac{1}{4}\boldsymbol{P}$$

we have

$$\begin{aligned}
&\boldsymbol{Q} + \boldsymbol{PQ} + \boldsymbol{P}^2\boldsymbol{Q} + \cdots \\
&= \left[\boldsymbol{I} + (\boldsymbol{P} + \boldsymbol{P}^2)\left(1 + \frac{1}{4} + \frac{1}{4^2} + \cdots\right)\right]\boldsymbol{Q} \\
&= \left[\boldsymbol{I} + \frac{4}{3}(\boldsymbol{P} + \boldsymbol{P}^2)\right]\boldsymbol{Q} \\
&= \begin{pmatrix}
\frac{1}{6} & \frac{1}{6} & \frac{1}{6} & \frac{1}{6} & \frac{1}{12} & \frac{1}{12} & \frac{1}{12} \\
\frac{7}{24} & \frac{1}{24} & \frac{1}{24} & \frac{7}{24} & \frac{13}{48} & \frac{1}{48} & \frac{1}{48} \\
\frac{7}{24} & \frac{7}{24} & \frac{1}{24} & \frac{1}{24} & \frac{13}{48} & \frac{1}{48} & \frac{1}{48} \\
\frac{1}{24} & \frac{7}{24} & \frac{7}{24} & \frac{1}{24} & \frac{1}{48} & \frac{13}{48} & \frac{1}{48} \\
\frac{1}{24} & \frac{1}{24} & \frac{7}{24} & \frac{7}{24} & \frac{1}{48} & \frac{1}{48} & \frac{13}{48}
\end{pmatrix}
\end{aligned}$$

It should be noted that the calculation program introduced here is faster than the usual program for solving the system of equations.

Now, let's look at the geometric meaning of the coefficients in equations (4) and (5). Looking at the coefficients $\frac{1}{6}$ and $\frac{1}{12}$ in equation (4), we can imagine: Starting from (0,0) to (0,2) there's a straight line, and to (1,1) there are two paths, "┌" and "┘" for a total of 12 paths. Therefore, the probability of reaching (0,2) is $\frac{1}{12}$, and the probability of reaching (1,1) is $\frac{2}{12}=\frac{1}{6}$, and so on.

This explanation is reasonable, but not easy to generalize. Here's an alternative explanation: The probability of moving from (0,0) to its east, south, west, or north neighboring points is each $\frac{1}{4}$. However, these four points are not boundary grid points, so the probability of reaching a boundary point from (0,0) in one step is zero.

The probability of reaching any interior point's four neighboring points is $\frac{1}{4}$. Therefore, taking two steps from (0,0), there are 16 possible outcomes. Among them, there is one path to each of the four vertices (4 total), two paths to each of the eight boundary points, and four ways to return to the origin. Thus, the probability of reaching each vertex in two steps is $\frac{1}{16}$, the probability of reaching each boundary point is $\frac{2}{16}=\frac{1}{8}$, and the probability of returning to the origin is $\frac{1}{4}$.

It's impossible to reach a boundary point from (0,0) in three steps.

Now, let's consider the case of taking four steps. The paths that reach the boundary in two steps are already accounted for. The next two steps are still starting from (0,0), but the probability of reaching the boundary points is multiplied by $\frac{1}{4}$. That is, the probability of reaching each vertex in four steps is $\frac{1}{4}\times\frac{1}{16}$ the probability of reaching each boundary point is $\frac{1}{4}\times\frac{1}{8}$, and the probability of returning to the origin is $\frac{1}{4}\times\frac{1}{4}$.

For five steps, it's not possible, but for six steps, the probabilities are

$$\frac{1}{4^2}\times\frac{1}{16},\ \frac{1}{4^2}\times\frac{1}{8},\ \frac{1}{4^2}\times\frac{1}{4}$$

Thus, the probability of reaching boundary points after an odd number of steps is zero. After $2l$ steps, the probabilities of reaching each vertex and boundary point are

$$\frac{1}{4^{l-1}}\times\frac{1}{16}$$

Probability of reaching each boundary point is

$$\frac{1}{4^{l-1}}\times\frac{1}{8}$$

Probability of returning to the origin is

$$\frac{1}{4^{l-1}}$$

Therefore, the probability of reaching each vertex is

$$\frac{1}{16}+\frac{1}{4}\times\frac{1}{16}+\frac{1}{4^2}\times\frac{1}{16}+\cdots=\frac{1}{16}\times\left(1+\frac{1}{4}+\frac{1}{4^2}+\cdots\right)=\frac{1}{16}\times\left(1-\frac{1}{4}\right)^{-1}=\frac{1}{12}$$

The probability of reaching each boundary point is

$$\frac{1}{8}\times\left(1+\frac{1}{4}+\frac{1}{4^2}+\cdots\right)=\frac{1}{6}$$

The probability of returning to the origin is

$$\lim_{l\to\infty}\frac{1}{4^l}=0$$

This describes random walks in probability theory.

Let's now look at the meaning of $\frac{13}{48}$ in equation (5): Starting from (1,0), the probability of reaching (2,0) in one step is $\frac{1}{4}$. The probability of moving from (1,0) in one step without reaching a boundary point is only possible to (0,0), with a probability of $\frac{1}{4}$, and subsequent cases are the same as starting from (0,0). Therefore, the probability of reaching (2,0) in more than one step is $\frac{1}{4}\cdot\frac{1}{12}$. The total probability is

$$\frac{1}{4}+\frac{1}{4}\cdot\frac{1}{12}=\frac{13}{48}$$

Similarly, the probability of reaching (1,1) or (1,−1) is

$$\frac{1}{4}+\frac{1}{4}\cdot\frac{1}{6}=\frac{7}{24}$$

Other boundary points must pass through (0,0), so the probability is the probability from (0,0) to those points multiplied by $\frac{1}{4}$, i.e.,

$$\frac{1}{4}\cdot\frac{1}{12}=\frac{1}{48},\ \frac{1}{4}\cdot\frac{1}{6}=\frac{1}{24}$$

From this simple example, it is intuitively clear how the general theory works. This also suggests that we use the probability method to solve the "Laplace equation boundary value problem." Essentially, it is solving the "differentiated algebraic system of equations."

2. Monte Carlo method (or probability method). Let's first define a two-dimensional random walk in general: Suppose there is a particle starting from a grid point in G^*, and it moves with equal probability 1/4 to its east, south, west, or north neighboring points. Then, from the new position, it moves again in the same way to one of the four neighboring points, and so on, until it reaches a boundary grid point, at which the walk terminates, as shown in figure 4:

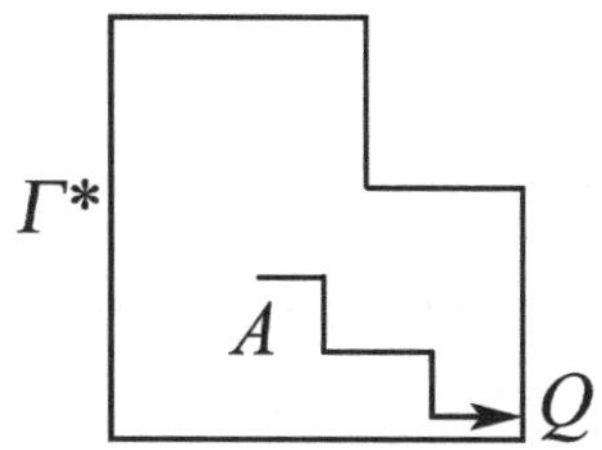

Figure 4

Let the random walk path be $\gamma_A: A \to A_1 \to \ldots \to A_{l-1} \to Q \in \Gamma^*$.

Then, we define the value of the random variable as $\xi = \xi(\gamma_A) = f(Q)$, where $f(Q)$ is the function value at the boundary point Q. If Q is a boundary point, then $\xi = \xi(\gamma_Q) = f(Q)$.

The expected value of the random variable ξ, denoted as $E(\xi)$, is the solution to the system of equations (2). In other words

$$E[\xi(\gamma_A)] = \frac{1}{4}\sum_{i=1}^{4} E[\xi(\gamma_{A_{1i}})],\ \text{if } A \in G^* \tag{9}$$

and

$$E[\xi(\gamma_Q)] = f(Q),\ \text{if } Q \in \Gamma^* \tag{10}$$

where A_{11}, A_{12}, A_{13}, A_{14} are the east, south, west, and north neighboring points of A, respectively.

Let $P(\gamma_A)$ denote the probability of following the path γ_A. Then

$$P(\Upsilon_A) = \frac{1}{4^l}$$

Thus

$$E[\xi(\gamma_A)] = \sum_{\gamma_A} \xi(\gamma_A) P(\gamma_A)$$

The sum on the right-hand side is over all possible paths starting from A. From A, the first step must be to one of its east, south, west, or north neighboring points $A_{11}, A_{12}, A_{13}, A_{14}$, and then the walk continues. Therefore,

$$E[\xi(\gamma_A)] = \sum_{i=1}^{4} \sum_{\gamma_{A_{1i}}} \xi(\gamma_{A_{1i}}) P(A \to A_{1i}) P(\gamma_{A_{1i}})$$

Since

$$P(A \to A_{1i}) = \frac{1}{4}$$

we get

$$\begin{aligned} E[\xi(\gamma_A)] &= \frac{1}{4} \sum_{i=1}^{4} \sum_{\gamma_{A_{1i}}} \xi(\gamma_{A_{1i}}) P(\gamma_{A_{1i}}) \\ &= \frac{1}{4} \sum_{i=1}^{4} E[\xi(\gamma_{A_{1i}})] \end{aligned}$$

This is equation (9). Similarly, when $Q \in \Gamma^{*}$, there is only one path, which is to stop at that point. Therefore,

$$E[\xi(\gamma_Q)] = \sum_{\gamma_Q} \xi(\gamma_Q) P(\gamma_Q) = \xi(\gamma_Q) = f(Q)$$

This gives equation (10).

Let $\xi_1, \ldots, \xi_N$ be the results of N observations of ξ. According to the law of large numbers, we know that for any $\varepsilon > 0$, there is

$$\lim_{N \to \infty} P\left(\left| E(\xi) - \frac{1}{N} \sum_{i=1}^{N} \xi_i \right| \leq \varepsilon \right) = 1$$

Thus, when N is large enough, the arithmetic mean

$$\frac{1}{N} \sum_{i=1}^{N} \xi_i$$

can be used as an approximation of $E(\xi)$ (i.e., the solution).

Random walks are typically implemented using physical methods or random numbers generated by mathematical methods. This explanation does not go into detail on this. However, a simple method is as follows: draw G^* on a Go board with chalk. To find the function value of a certain point, mark it, place a piece on the board, and then roll a tetrahedral die labeled with east, south, west, and north (as shown in fig. 5). If the die lands with the east face (or south, west, or north), move one step in that direction, then roll the die again and continue until a boundary is reached. This way, we obtain a random walk, and the function value at the boundary point is the value of the random variable ξ. By performing many walks (let's say N times), recording $\xi_1, \ldots, \xi_N$, the arithmetic mean will give the desired approximate function value.

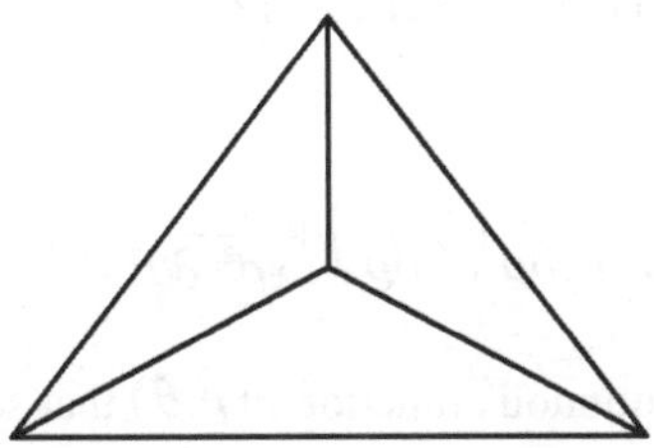

Figure 5

Conclusion: The error in the finite difference method consists of three parts: 1. The error introduced when discretizing the grid boundary values. 2. The error introduced when replacing derivatives with differences during discretization. 3. The errors introduced by solving the finite difference equations—ordinary errors in the algebraic method and probabilistic errors in the Monte Carlo method.

Therefore, the error produced by the Monte Carlo method is larger and less reliable than the error produced by the algebraic method. However, on the other hand, the Monte Carlo method has a very simple calculation program. Additionally, if we only need to find the function values at certain grid points and not at all points, this method becomes more economical.

■ Analytical Expressions: Sometimes Leading to Misguidance

Some analytical formulas may seem promising, appearing to effectively solve a problem, even potentially offering a complete solution. However, if applied thoughtlessly, they may lead us down a misguided path. By thoroughly understanding the relationship between "continuity" and "discreteness," these mistakes can be completely avoided! Moreover, in contrast, there is a complementary relationship, where analytical expressions can spark new computational methods, not only for their theoretical importance.

Let's continue with the example of the Dirichlet problem for the Laplace equation, assuming the region is the unit circle. The polar form of the Laplace equation is

$$\frac{1}{\rho}\frac{\partial}{\partial\rho}\left(\rho\frac{\partial u}{\partial\rho}\right)+\frac{1}{\rho^2}\frac{\partial^2 u}{\partial\theta^2}=0 \tag{1}$$

Problem: Find the continuous function $u(\rho,\theta)$ that satisfies equation (1) inside the unit circle and matches a given continuous function on the boundary Γ, i.e.,

$$u(\rho,\theta)|_{\rho=1}=\varphi(\theta) \tag{2}$$

Assume $\varphi(\theta)$ has continuous derivatives up to order r($\geq$2) in [0,2π], and it is a periodic function with a period of 2π, where $|\varphi^{(r)}(\theta)|< C$. Expand $\varphi(\theta)$ as a Fourier series

$$\varphi(\theta)=\frac{a_0}{2}+\sum_{m=1}^{\infty}(a_m\cos m\theta+b_m\sin m\theta) \tag{3}$$

where

$$a_m=\frac{1}{\pi}\int_0^{2\pi}\varphi(\theta)\cos m\theta d\theta$$
$$b_m=\frac{1}{\pi}\int_0^{2\pi}\varphi(\theta)\sin m\theta d\theta \tag{4}$$

It is clear that

$$\rho^m\cos m\theta,\ \rho^m\sin m\theta(m=0,1,2,\dots) \tag{5}$$

are solutions to equation (1) with boundary values given by $\cos m\theta$ and $\sin m\theta$, respectively. Therefore, we can expect that

$$u(\rho,\theta)=\frac{a_0}{2}+\sum_{m=1}^{\infty}(a_m\cos m\theta+b_m\sin m\theta)\rho^m \tag{6}$$

is the solution to equation (1) that satisfies both equation (2) and equation (3). Given that

$$a_m=O\left(\frac{1}{m^r}\right),\ b_m=O\left(\frac{1}{m^r}\right)$$

equation (6) is indeed a solution to equation (1) that satisfies both conditions (2) and (3). Since

$$\begin{aligned}\frac{1}{2}+\sum_{m=1}^{\infty}\rho^m\cos m\theta &= R\left(\frac{1}{1-\rho e^{i\theta}}\right)-\frac{1}{2}\\ &=\frac{1-\rho\cos\theta}{1-2\rho\cos\theta+\rho^2}-\frac{1}{2}\\ &=\frac{1-\rho^2}{2(1-2\rho\cos\theta+\rho^2)}\end{aligned}$$

we obtain from equation (4) and equation (6):

$$\begin{aligned}u(\rho,\theta)&=\frac{1}{\pi}\int_0^{2\pi}\varphi(\psi)\left[\frac{1}{2}+\sum_{m=1}^{\infty}(\cos m\theta\cos m\psi+\sin m\theta\sin m\psi)\rho^m\right]\mathrm{d}\psi\\ &=\frac{1}{\pi}\int_0^{2\pi}\varphi(\psi)\left[\frac{1}{2}+\sum_{m=1}^{\infty}\rho^m\cos m(\theta-\psi)\right]\mathrm{d}\psi\\ &=\frac{1}{2\pi}\int_0^{2\pi}\varphi(\psi)\frac{1-\rho^2}{1-2\rho\cos(\theta-\psi)+\rho^2}\mathrm{d}\psi\end{aligned}\tag{7}$$

This is the Poisson formula.

This is the analytical solution to $u(\rho,\theta)$. This formula is indeed quite promising and seems to fully solve the problem. But upon closer inspection, we must ask: Does it really solve the problem?

If $\varphi(\psi)$ is given, we could compute the integral in equation (7) (i.e., find the original function), and the problem would be completely solved. However, if the integral cannot be computed (which happens much more often than the case where it can), or if the boundary values are given only as a set of data from experiments, the problem arises of how to approximate the solution $u(\rho,\theta)$. Naturally, one might consider using numerical integration to approximate the solution to equation (7). However, as we will see below, this approach leads to absurd conclusions.

1. Using the rectangular rule, we approximate $u(\rho,\theta)$ with

$$T_n(\rho,\theta)=\frac{1}{n}\sum_{l=0}^{n-1}\varphi\left(\frac{2\pi l}{n}\right)\frac{1-\rho^2}{1-2\rho\cos\left(\theta-\frac{2\pi l}{n}\right)+\rho^2} \tag{8}$$

Now, consider the behavior when $\rho\to 1-0$,

$$\lim_{\rho\to 1-0}T_n(\rho,\theta)=\begin{cases}0,\ \text{when}\ \theta\neq\frac{2\pi l}{n},\ \text{or}\ \theta=\frac{2\pi l}{n}\ \text{and}\ \varphi\left(\frac{2\pi l}{n}\right)=0\ (0\le l<n)\\ \infty,\ \text{when}\ \theta=\frac{2\pi l}{n}\ \text{and}\ \varphi\left(\frac{2\pi l}{n}\right)\neq 0\ (0\le l<n)\end{cases} \tag{9}$$

Thus, using $T_n(\rho,\theta)$ to approximate $u(\rho,\theta)$ is utterly absurd.

2. In the Poisson integral, we use a step function:

$$\varphi^*(\theta)=\varphi\left(\frac{2\pi l}{n}\right)$$

where

$$\frac{2\pi l}{n}\le\theta<\frac{2\pi(l+1)}{n}\quad(0\le l<n) \tag{10}$$

substitutes $\varphi(\theta)$.

In other words, we approximate $u(\rho,\theta)$ using

$$R_n(\rho,\theta)=\frac{1}{2\pi}\int_0^{2\pi}\frac{\varphi^*(\psi)(1-\rho^2)}{1-2\rho\cos(\psi-\theta)+\rho^2}d\psi \tag{11}$$

Since

$$\log(1-\rho e^{i\theta})=-\sum_{m=1}^{\infty}\frac{(\rho e^{i\theta})^m}{m}$$

taking the imaginary part yields

$$\sum_{m=1}^{\infty} \rho^m \frac{\sin m\theta}{m} = \operatorname{tg}^{-1} \frac{\rho \sin \theta}{1 - \rho \cos \theta}$$

Thus,

$$\begin{aligned} R_n(\rho,\theta) &= \sum_{l=0}^{n-1} \frac{\varphi\left(\frac{2\pi l}{n}\right)}{2\pi} \int_{\frac{2\pi l}{n}}^{\frac{2\pi(l+1)}{n}} \frac{1-\rho^2}{1-2\rho\cos(\theta-\psi)+\rho^2} \mathrm{d}\psi \\ &= \frac{1}{n}\sum_{l=0}^{n-1} \varphi\left(\frac{2\pi l}{n}\right) + \frac{1}{\pi}\sum_{l=0}^{n-1} \varphi\left(\frac{2\pi l}{n}\right) \cdot \\ &\quad \sum_{m=1}^{\infty} \left[\sin m\left(\frac{2\pi(l+1)}{n} - \theta\right) - \sin m\left(\frac{2\pi l}{n} - \theta\right)\right] \frac{\rho^m}{m} \qquad (12) \\ &= \frac{1}{n}\sum_{l=0}^{n-1} \varphi\left(\frac{2\pi l}{n}\right) + \frac{1}{\pi} \\ &\quad \sum_{l=0}^{n-1} \left[\varphi\left(\frac{2\pi(l+1)}{n}\right) - \varphi\left(\frac{2\pi l}{n}\right)\right] \operatorname{tg}^{-1} \frac{\rho \sin\left(\frac{2\pi l}{n} - \theta\right)}{1 - \rho\cos\left(\frac{2\pi l}{n} - \theta\right)} \end{aligned}$$

Let

$$\theta = C(1-\rho)^{\alpha} \left(\alpha > \frac{1}{2}\right)$$

Then,

$$\lim_{\rho \to 1-0} \operatorname{tg}^{-1} \frac{\rho \sin\theta}{1-\rho\cos\theta} = \lim_{\rho \to 1-0} \operatorname{tg}^{-1} C(1-\rho)^{\alpha-1}$$

In other words, as $\rho \to 1-0, \theta \to 0$, the value of

$$\operatorname{tg}^{-1} \frac{\rho \sin\theta}{1-\rho\cos\theta}$$

can approach any value in the interval

$$\left[-\frac{\pi}{2}, \frac{\pi}{2}\right]$$

Therefore, if

$$\theta_0 = \frac{2\pi l}{n}$$

and

$$\varphi\left[\frac{2\pi(l-1)}{n}\right] \neq \varphi\left(\frac{2\pi l}{n}\right)(0 \leq l < n)$$

then as $\rho \to 1-0, \theta \to \theta_0$, the limit of $R_n(\rho,\theta)$ does not exist. Therefore, the method of limit approach must be restricted, for example, by specifying that the limit is taken along the radial direction, etc. Furthermore, it can be proven that even with these restrictions, using $R_n(\rho,\theta)$ to approximate $u(\rho,\theta)$ is still not highly precise. A detailed discussion is omitted here.

Both of the approximation methods derived from the analytical formula above are less precise than the elementary method below:

Given the function values at $n(=2n'+1)$ points,

$$y_l = \varphi\left(\frac{2\pi l}{n}\right)(|l| \leq n')$$

Then, as shown in section 4, define

$$S_n(\theta) = \sum_{m=-n'}^{n'} C'_m \mathrm{e}^{\mathrm{i}m\theta}$$

$$C'_m = \frac{1}{n}\sum_{l=-n'}^{n'} y_l \mathrm{e}^{-2\pi\mathrm{i}lm/n} \tag{13}$$

If $\varphi(\theta)$ has r(≥2) continuous derivatives in $[-\pi,\pi]$, is a periodic function with a period of 2π, and satisfies $|\varphi^{(r)}(\theta)| < C$ then

$$|\varphi(\theta) - S_n(\theta)| < \frac{4C}{(r-1)n'^{r-1}} \tag{14}$$

Define

$$S_n(\rho,\theta) = \sum_{m=-n'}^{n'} C'_m \mathrm{e}^{\mathrm{i}m\theta}\rho^{|m|} \tag{15}$$

Then

$$u(\rho,\theta)-S_n(\rho,\theta)=\frac{1}{2\pi}\int_{-\pi}^{\pi}\frac{[\varphi(\psi)-S_n(\psi)](1-\rho^2)}{1-2\rho\cos(\theta-\psi)+\rho^2}\mathrm{d}\psi$$

Thus,

$$\begin{aligned}&|u(\rho,\theta)-S_n(\rho,\theta)|\\&\leq\frac{1}{2\pi}\int_{-\pi}^{\pi}\frac{|\varphi(\psi)-S_n(\psi)|(1-\rho^2)}{1-2\rho\cos(\theta-\psi)+\rho^2}\mathrm{d}\psi\\&<\frac{4C}{(r-1)n'^{r-1}}\cdot\frac{1}{2\pi}\int_{-\pi}^{\pi}\frac{1-\rho^2}{1-2\rho\cos(\theta-\psi)+\rho^2}\mathrm{d}\psi\\&=\frac{4C}{(r-1)n'^{r-1}}\end{aligned}\tag{16}$$

In practical calculations, because

$$\begin{aligned}\sum_{m=-l}^{l}\mathrm{e}^{\mathrm{i}mx}\rho^{|m|}&=\sum_{m=0}^{l}(\mathrm{e}^{\mathrm{i}x}\rho)^m+\sum_{m=0}^{l}(\mathrm{e}^{-\mathrm{i}x}\rho)^m-1\\&=\frac{1-\mathrm{e}^{\mathrm{i}(l+1)x}\rho^{l+1}}{1-\rho\mathrm{e}^{\mathrm{i}x}}+\frac{1-\mathrm{e}^{-\mathrm{i}(l+1)x}\rho^{l+1}}{1-\rho\mathrm{e}^{-\mathrm{i}x}}-1\\&=\frac{2-2\rho^{l+1}\cos(l+1)x-2\rho\cos x+2\rho^{l+2}\cos lx}{1-2\rho\cos x+\rho^2}-1\\&=\frac{1-\rho^2-2\rho^{l+1}\cos(l+1)x+2\rho^{l+2}\cos lx}{1-2\rho\cos x+\rho^2}\end{aligned}$$

then

$$\begin{aligned}S_n(\rho,\theta)&=\sum_{m=-n'}^{n'}\frac{1}{n}\sum_{l=-n'}^{n'}y_l\mathrm{e}^{-2\pi\mathrm{i}lm/n}\mathrm{e}^{\mathrm{i}m\theta}\rho^{|m|}\\&=\frac{1}{n}\sum_{l=-n'}^{n'}y_l\sum_{m=-n'}^{n'}\mathrm{e}^{\mathrm{i}\left(\theta-\frac{2\pi l}{n}\right)m}\rho^{|m|}\end{aligned}$$

$$=\frac{1}{n}\sum_{l=-n'}^{n'} y_l \frac{\left\{\begin{matrix} 1-\rho^2-2\rho^{n'+1}\cos(n'+1)\left(\theta-\frac{2\pi l}{n}\right) \\ +2\rho^{n'+2}\cos n'\left(\theta-\frac{2\pi l}{n}\right)\end{matrix}\right\}}{1-2\rho\cos\left(\theta-\frac{2\pi l}{n}\right)+\rho^2} \tag{17}$$

In conclusion, we obtain **Theorem 1**: Let $u(\rho,\theta)$ be the solution to equation (1) that satisfies equation (2), where $\varphi(\theta)$ is a function with r(≥2) continuous derivatives and has a period of 2π, and it is assumed that $|\varphi^{(r)}(\theta)|<C$. Then

$$\left| u(\rho,\theta)-\frac{1}{n}\times\sum_{l=-n'}^{n'}\varphi\left(\frac{2\pi l}{n}\right) \left[\frac{1-\rho^2-2\rho^{n'+1}\cos(n'+1)\left(\theta-\frac{2\pi l}{n}\right)}{1-2\rho\cos\left(\theta-\frac{2\pi l}{n}\right)+\rho^2}+\frac{2\rho^{n'+2}\cos n'\left(\theta-\frac{2\pi l}{n}\right)}{1-2\rho\cos\left(\theta-\frac{2\pi l}{n}\right)+\rho^2}\right]\right| < \frac{4C}{(r-1)n'^{r-1}} \tag{18}$$

■ Uniform Distribution: Number Theory Methods and Monte Carlo Methods

To calculate the integral of a function *f(x)* over [0,1], we can divide [0,1] into *n* equal parts, take the arithmetic average of the function values at the partition points, and use it as an approximation of the integral of *f(x)* (the rectangle rule). This is the method of converting continuous into discrete. In fact, not only the equidistant points have this property, but any sequence of points that satisfies the so-called "uniform distribution" condition also has this property. Roughly speaking, the meaning of uniform distribution is that the likelihood of the points being near any point in [0,1] is equal. Strictly speaking, it can be defined as follows:

Let $x_i (i = 1, 2, \ldots)$ be a sequence of points in [0,1], and let a be any real number such that $0 \leq a \leq 1$. Let $N_n(a)$ denote the number of points $x_1, \ldots, x_n$ that fall in the subinterval [0,a]. If the following condition holds

$$\lim_{n \to \infty} \frac{N_n(a)}{n} = a \tag{1}$$

then the sequence $x_i (i = 1, 2, \ldots)$ is said to be uniformly distributed in [0,1].

The following is a criterion for uniform distribution.

Theorem 1: The necessary and sufficient condition for the sequence of points

$$x_1, \ldots, x_m, \ldots, 0 \leq x_m \leq 1 \tag{2}$$

to be uniformly distributed is that for any function *f(x)* that is Riemann integrable over [0,1], the following condition holds

$$\lim_{n \to \infty} \frac{f(x_1) + \ldots + f(x_n)}{n} = \int_0^1 f(x)\mathrm{d}x \tag{3}$$

Proof: We first prove that if $\{x_i\}$ is uniformly distributed, then equation (3) holds.

1. Let *f(x)* be a function defined as

$$f(x) = \begin{cases} C, \text{if } 0 \leq x \leq a \\ 0, \text{otherwise} \end{cases}$$

Then

$$\lim_{n \to \infty} \frac{f(x_1) + \ldots + f(x_n)}{n} = C \lim_{n \to \infty} \frac{N_n(a)}{n} = Ca = \int_0^1 f(x)\mathrm{d}x$$

Thus, for such a function *f(x)*, the theorem is true.

2. If equation (3) holds for functions $f_1, \ldots, f_s$, then it holds for any linear combination $c_1 f_1 + \ldots + c_s f_s$, and thus equation (3) is true for all step functions.

3. It is well-known that if f is Riemann integrable, then for any $\varepsilon > 0$, there exist two step functions $\varphi_\varepsilon(x)$ and $\Phi_\varepsilon(x)$ such that

$$\varphi_\varepsilon(x) \leq f(x) \leq \Phi_\varepsilon(x), \ 0 \leq x \leq 1 \tag{4}$$

and

$$\int_0^1 [\Phi_\varepsilon(x) - \varphi_\varepsilon(x)]\mathrm{d}x < \varepsilon \tag{5}$$

From 2. we know that the theorem holds for $\varphi_\varepsilon(x)$ and $\Phi_\varepsilon(x)$, then

$$\begin{aligned}\int_0^1 \varphi_\varepsilon(x)\mathrm{d}x &= \lim_{n\to\infty}\frac{1}{n}[\varphi_\varepsilon(x_1)+\ldots+\varphi_\varepsilon(x_n)] \\ &\le \lim_{n\to\infty}\frac{1}{n}[f(x_1)+\ldots+f(x_n)] \\ &\le \lim_{n\to\infty}\frac{1}{n}[\Phi_\varepsilon(x_1)+\ldots+\Phi_\varepsilon(x_n)] \\ &= \int_0^1 \Phi_\varepsilon(x)\mathrm{d}x\end{aligned}$$

So, we have

$$\left|\lim_{n\to\infty}\frac{f(x_1)+\ldots+f(x_n)}{n} - \int_0^1 f(x)\mathrm{d}x\right| < \varepsilon$$

This proves the necessary part of the theorem.

The proof of the sufficient part is straightforward: simply take

$$f(x)=\begin{cases}1, \text{if } 0\le x\le a\\ 0, \text{otherwise}\end{cases}$$

Then equation (3) becomes

$$\lim_{n\to\infty}\frac{N_n(a)}{n} = a$$

The theorem is thus proved.

Clearly, the definition of uniform distribution and its criterion can be easily extended to multivariable cases (i.e., to the unit cube in higher dimensions). According to theorem 1, numerical integration methods essentially depend on the selection of uniformly distributed point sequences. The central problem in numerical integration is how to select the best uniformly distributed sequence of points.

It is well known that for computing integrals over [0,1], using equally spaced points yields the most accurate error (in terms of the order of the error). However, in the multivariable case, if equally spaced points are used for computation, the error depends on the number of variables (i.e., the dimension of the integral). More precisely, if the number of sampling points is fixed, then as the number of variables increases, the error also increases rapidly. In other words, to achieve a certain level of accuracy, the number of required sampling points must grow rapidly with the number of variables. Therefore, using this method to handle numerical integration in high-dimensional spaces leads to a massive amount of computation and becomes practically infeasible. Specifically, for an s-dimensional integral, to achieve an error of order $O(1/n)$, the number of sampling points needed is $O(n^s)$.

In recent years, the Monte Carlo method has been developed as a widely used method for numerical integration in high-dimensional spaces. This method involves randomly selecting n points $\left(x_1^{(k)},\ldots,x_s^{(k)}\right)(k=1,2,\ldots,n)$, and then using the arithmetic average of the function values at these n points to approximate the integral. The term "randomly" means that each point is selected with equal probability. When n is sufficiently large, this method can achieve a certain level of accuracy. The process of selecting random points is usually carried out on a computer using mathematical methods, many of which are number-theoretic methods, especially those involving congruences. The advantage of the Monte Carlo method is its simplicity in computational implementation. Although its convergence rate is faster than the rectangle rule, the error obtained by this method is a probabilistic error, not an exact error.

The so-called number-theoretic method approximates multidimensional integrals by taking the sum of function values at preselected optimally distributed point sequences. Therefore, the error is deterministic rather than probabilistic. Moreover, these deterministic errors are often better than the probabilistic errors. In fact, it can be proven that for certain classes of functions, the leading term of the approximation error has already reached the optimal level. Specifically, the leading order of the error is the same as that for one-dimensional integrals.

Example: Assume that $f(x_1,\ldots,x_5)$ is a function that has second-order continuous partial derivatives with respect to each variable, and that all its partial derivatives are periodic functions of each variable with period 1, and satisfy

$$\left|\frac{\partial^r f}{\partial x_1^{i_1}\cdots\partial x_5^{i_5}}\right| < C(2\pi)^r\,(i_1+\cdots+i_5=r, 0\le r\le 10, 0\le i_j\le 2)$$

Then

$$\left|\int_0^1 \cdots \int_0^1 f(x_1, \ldots, x_5)dx_1 \ldots dx_5 - \frac{1}{15019}\right.$$
$$\left.\sum_{k=1}^{15019} f\left(\frac{k}{15019}, \frac{10641k}{15019}, \frac{2640k}{15019}, \frac{6710k}{15019}, \frac{781k}{15019}\right)\right|$$
$$< 0.0032\left(\frac{\pi^2}{6}\right)^5 C$$

It must be pointed out that number-theoretic methods are not only useful in numerical integration, but also applicable in fields such as function approximation theory and the asymptotic solution of integral equations. For example, it can be proven that multivariable functions satisfying certain smoothness and periodicity conditions (with period 1 in each variable) can be approximated by a trigonometric polynomial, and the leading order of the approximation error is independent of the number of variables. For more on this, see [1].

■ References

1. Hua Luogeng, Wang Yuan. Numerical integration and its applications. Beijing: Science Press; 1963.
2. Hua Luogeng. Introduction to advanced mathematics. Beijing: Science Press; 1963.

Time—Oh Time—Never to Return*

■ Time and Space

People love to talk about four-dimensional (or four-degree) space: front-back, left-right, and up-down are the three dimensions related to spatial position; past, present, and future make up the one dimension related to time. Broadly speaking, we live in a four-dimensional world, where every material object occupies a specific position at a specific time. Although time and space are both referred to as "dimensions," their nature is vastly different.

If you walk three miles in the wrong direction, it's no big deal—you can walk back and return to your starting point. But time cannot be turned back to the original moment—you just end up spending double the time. Time—oh time—never to return! The "present" is always a fleeting instant that vanishes in the blink of an eye. A person only lives once—there's no room for carelessness. As the old poem aptly puts it: "A hundred streams flow eastward to the ocean, / Nevermore to turn west again. / One who mis-spends his youth, / In old age will grieve in vain."

However, if one compares the passage of time to flowing water, there is no true cycle of return. Water, after tumbling through rivers and oceans, turns into vapor, forms clouds, rides the wind, becomes rain—and may occasionally return to the highlands. Only time possesses the true quality of once gone, never returning.

---

* First published in *People's Daily* on March 30, 1964.

■ A Matter of Life Philosophy

How should one spend and make use of their spare time? It may seem like a question of personal interest, but in fact, it is a matter of life philosophy—a question of stance.

Since time flies and youth doesn't last forever, why not enjoy life while we can? That is one type of life philosophy—a decadent, degenerate bourgeois philosophy. For someone with this kind of life philosophy, not only will they fail to make good use of their spare time, but even during work hours, they will be careless and lazy, working only for the sake of a paycheck. When under the supervision of leaders or colleagues, they might reluctantly put in more effort—but otherwise, they slack off and daydream, thinking, "There's always tomorrow."

But for a revolutionary, their entire life belongs to the Party and the people. Not only will they strive to make every second of work time more effective in service of the revolutionary cause, but even in their spare time, they will take what benefits the revolution as their guiding principle. Even when working alone, they will "practice self-discipline," not merely avoiding wrongdoing but actively and wholeheartedly striving to do things that benefit the people, seizing every moment. Exercise is for health—so one can shoulder more demanding work for the people and extend one's years of effective service. Cultural and recreational activities are for relieving fatigue—so one can work with more energy, greater spirit, sharper thought, and stronger focus. Even socializing with friends is based on mutual help and shared growth.

The same reality—that time never returns—can lead to two vastly different approaches: One treats work as a means for rest and enjoyment; and the other treats rest as a means to work better and more effectively.

■ Interests Can Be Cultivated

Once a person has adopted the stance and determination to serve the people, everything gains a purpose—they will take every opportunity to train and develop themselves to better meet the needs of socialist society. Political study, physical exercise, professional training, cultural entertainment—all these will be arranged with a clear and correct goal. Over time, these efforts will become habits, and one will find joy and fulfillment in them.

I have a little personal experience: Before liberation, I enjoyed writing academic papers and had never written a single popular science article. But under the Party's education—especially after studying Chairman Mao's "Talks at the Yan'an Forum on Literature and Art"—I came to understand that the mission of a scientific worker should include not only advancing knowledge, but also popularizing it. We should write educational materials for junior high school students. So I forced myself to try writing such articles. Of course, I still haven't written them very well even now—but I've developed an interest, and I enjoy writing and reading this kind of work, and even analyzing it.

Of course, this doesn't mean we should reject hobbies that develop personal talents. On the contrary, such hobbies, when properly arranged, are not a waste of time at all—they provide balance and enrichment.

■ Who Only Worked Eight Hours a Day Before Liberation?

In the old society, who only worked eight hours a day? Some did—some even worked only two or three hours, or didn't work at all, indulging in leisure as they pleased. They had absolute control over all 24 hours of their day, year-round: "In spring, wandering through fields of fragrant grass; in summer, admiring the green lotus ponds; in autumn, drinking chrysanthemum wine; in winter, composing poems about the white snow." These were the so-called "refined scholars of leisure." Others were even worse, spending entire nights playing mahjong, indulging in all manner of vices—drinking, womanizing, and gambling—engaging in every conceivable depravity. This was the lifestyle of the landlord class, the idle class, the exploiting class.

But for the common laboring people, who could afford to work only eight hours a day? They toiled endlessly in the fields, factories, mines, rivers, docks, and shops, working themselves to the bone! Yet, even with such exhausting labor, they still struggled to make ends meet. In reality, the reason why some people in the old society could live without working was precisely because others were forced to work themselves to exhaustion—10 hours, 12 hours, even 16 hours a day. That was the essence of exploitation. The idea of making others work more while doing less oneself is, at its core, rooted in an exploitative mindset.

In the new society, where there is no exploitation, everyone strives to work to the best of their abilities, competing to be more efficient and contributing to the collective good.

Even in capitalist countries, you hear about "six-hour workdays" or even "four-hour workdays." But in reality, what happens? A job meant for one person is split between two, reducing unemployment numbers artificially. Yet workers are left with half the wages, struggling to survive on meager earnings. This is just another trick of the capitalist class to manage unemployment.

■ Every Second Adds Up Over Time

It's not just about how one spends large blocks of free time—learning to use scattered moments effectively is just as important. For instance, if a meeting is scheduled for 9:00 a.m., some people waste their entire morning waiting for it, unable to focus on anything else. If there are two hours of class in the morning, they let the time before and after slip away without doing anything productive. This is highly inefficient. Time is made up of seconds and minutes—those who know how to utilize these fragments of time will accomplish much more.

While valuing one's own time is important, respecting other people's time is even more so. If a teacher spends an extra hour preparing a lesson, it might save every student in the class an hour of confusion. That is a worthwhile investment of time. Within a collective, we should be eager to take on difficult tasks and act as a "stepping stone" for others. For example, if someone is engaged in an important project, we should be willing to take on other responsibilities for them, allowing them to concentrate fully and complete their work more efficiently. This is the essence of socialist teamwork and collaboration, completely different from the jealousy and sabotage seen in the old society, where people would rather see others fail if they themselves couldn't succeed.

From a societal perspective, time is a collective resource, yet it is composed of the individual seconds and minutes contributed by each person. We shouldn't only focus on utilizing our own time efficiently—we should also consider the bigger picture and help others use their time more effectively. This applies to leisure time as well. Idle gossip, meaningless chatter (especially not about politics and expertise), and spreading rumors waste time and sow discord—this is something to be avoided. On the other hand, discussing ideas, exchanging knowledge, and sharing experiences foster learning and mutual improvement—this is a practice worth encouraging. The former weakens collective strength, while the latter strengthens it—making a world of difference for socialism.

In short, leisure time is just as precious as working hours. But how one chooses to use it depends on the heart—on a wholehearted commitment to serving the people.

Popular Explanation of Optimization and Its Supplement

■ Popular Explanation of Optimization

► What Is Optimization?

Optimization problems are everywhere, but since many of them are simple and easy to solve, they tend not to attract much attention. However, as industrial processes have grown increasingly complex and demands for higher quality have become stricter, optimization problems have gained more prominence.

Let's start with a simple example: What is the best length for a piece of chalk? Each stick of chalk ends up with a leftover stub that gets discarded. From this standpoint, the longer the chalk, the better, since the waste ratio is lower. But if the chalk is too long, it becomes unwieldy and more likely to break. And every time it breaks, it results in extra waste. So, there's a trade-off. This leads to the problem: What is the optimal chalk length?

Another example: How much baking soda should be used when steaming buns? Too much and the buns taste bad; too little and they taste bad as well. So, what's the right amount for the best taste? Again, this is an optimization problem.

Someone might argue: This is an imprecise question because "taste" is subjective—I like it one way, you like it another—there's no objective standard for what's tasty. That's a fair point—but not entirely accurate. Can't we create a standard tailored to our own dining hall, for example? Suppose there are 100 people in the cafeteria. For different amounts of baking soda used, we could count how many people think the buns taste good. That number becomes our metric. Then our optimization problem becomes: Find the amount of baking soda that makes the most people in the cafeteria say the buns taste good.

This is just an analogy—a way to explain the idea simply. In reality, things are more complex: Issues like fermentation also play a role, and we haven't considered those here. Besides, seasoned cooks have already figured this out through long practice. We're only using this example to help explain what optimization means.

The optimization method applies in situations like the following: How to select the best formulation or production process to achieve the highest product quality? How to achieve the highest output, lowest cost, or fastest production while meeting quality standards? How to tune existing instruments to achieve the best performance?

Someone might say: Why not just run a large number of experiments? Try every possible option—surely we'll find the best one! But conducting extensive experiments consumes huge amounts of time, effort, and materials—and sometimes it's not even feasible. Here's a simple example: Imagine a pond that covers one square kilometer, and we want to find the deepest point. Suppose we decide to measure the depth every meter—that's a grid of 1,000 × 1,000 = one million measurement points (fig. 1). And that's not complex for there are just two variables—the horizontal X and Y directions. What if there is more—three, four, five, or six variables? It gets overwhelming quickly.

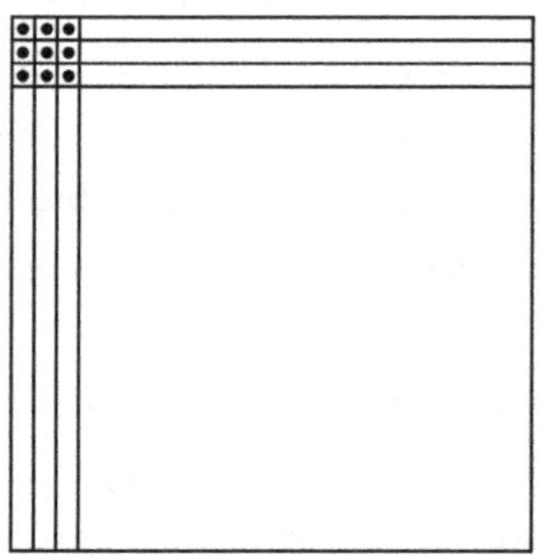

Figure 1

Let's say we need 100 levels of precision per variable (i.e., each factor is divided into 100 grades). 2 factors = 100 × 100 = 10,000 trials; 3 factors = 100^3 =

1,000,000 trials; and 4 factors = 100^4 = 100 million trials. Even if we could do 30 trials a day, or 10,000 per year, it would still take 10,000 years to finish.

The goal of optimization is to reduce the number of experiments needed while still finding the optimal solution. For example, with one factor, optimization might need just 14 experiments instead of 1,600. For the pond example, we might only need 130 samples instead of 1 million, provided the pond bottom isn't drastically irregular.

▸ Single-Factor Optimization

We know that certain chemical elements must be added to steel to enhance its strength. Too little is ineffective, and too much is also detrimental. For example, if the carbon content is too high, it becomes cast iron; if too low, it becomes wrought iron—neither qualifies as proper steel. So, how much carbon should be added per ton to achieve maximum strength? Suppose it's estimated (or theoretically calculated) that the optimal range is between 1,000 grams and 2,000 grams per ton.

The conventional method would be to test at 1,001 grams, 1,002 grams, and so on, all the way to 2,000 grams. After a thousand tests, the best option might be found. This method is called the equal interval method. Performing a thousand tests not only wastes time and effort but also consumes a significant amount of raw materials. To rapidly determine the optimal solution, we suggest the following "folded strip method."

Please remember this number: 0.618.

Use a strip of paper marked with a scale representing 1,000 to 2,000 grams (fig. 2). Draw a line at the 0.618 point of this strip (fig. 3). Conduct a test at the amount indicated by this line—i.e., 1,618 grams.

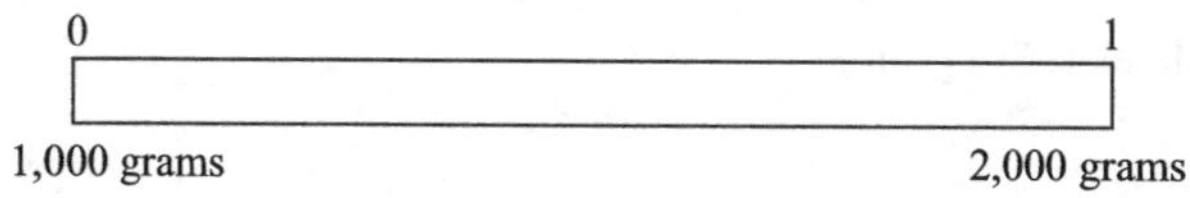

Figure 2

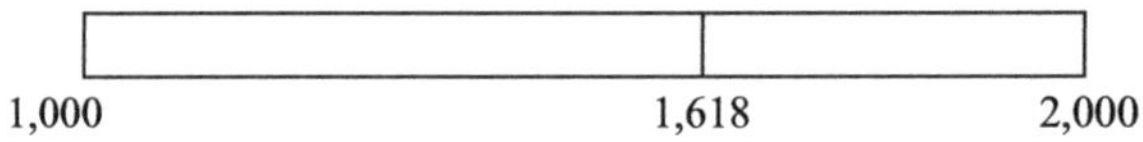

Figure 3

Next, fold the strip in half so that the previous mark aligns with the other side of the strip. Draw a new line where the fold intersects the scale (fig. 4), which falls at 1,382 grams. Conduct a second test at this amount.

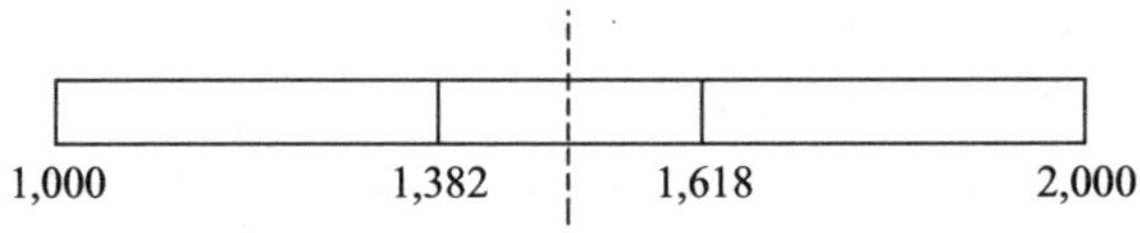

Figure 4

Now compare the results of the two tests. If 1,382 grams performs better, cut off the right side of the strip at 1,618 grams (fig. 5). (If 1,618 grams performs better, cut off the left side at 1,382 grams instead.)

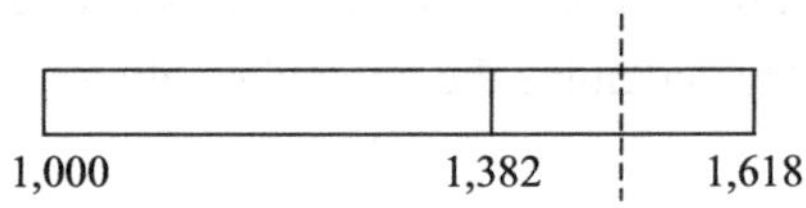

Figure 5

Fold the remaining strip in half again to find a new test point—1,236 grams (fig. 6). Test this amount and compare with 1,382 grams. If 1,382 grams is still better, cut off the left side at 1,236 grams.

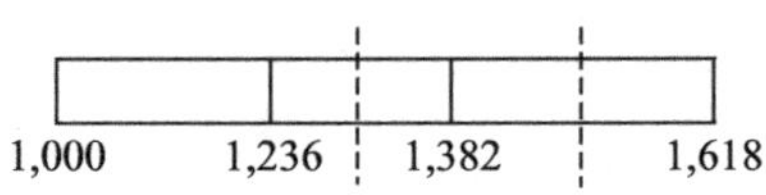

Figure 6

Repeat the process: fold, draw a new test point (e.g., 1,472 grams, fig. 7), test, and cut the strip accordingly. Note that each time, the remaining length of the strip is 0.618 times the previous length (remaining length = 0.618 × previous length).

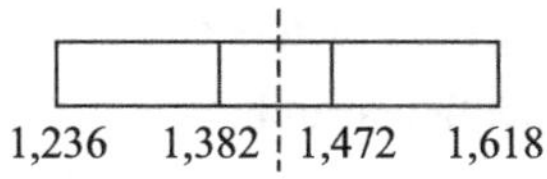

Figure 7

In this way—through testing, analyzing, testing again, and further analyzing—each round brings us closer to the desired amount, resolving contradictions as they arise, and gradually approaching the optimal value with increasing accuracy, until the required level of precision is achieved.

The history of steelmaking clearly illustrates the importance of optimization. Early iron production yielded cast iron with about 4% carbon. Later came wrought iron with almost no carbon. Until the 1770s in Europe, wrought iron was still widely used. The emergence of different types of steel reflects the process of finding the most appropriate carbon content to meet various practical needs: low-carbon steel (0.15% C) for stamping car bodies; mild steel (0.25% C) for large I-beams in construction; medium-carbon steel (0.5% C), heat-treated for axles and shafts; and high-carbon steel (1.4% C) for springs, hammers, files, and axes. For alloy steels, selecting the right composition becomes even more critical.

Steel is just one example. In fields like chemical industries with complex formulations and electronics industries with intricate production conditions, the need for optimization is even greater.

Grasping the Principal Contradiction

Things are complex and determined by many factors, so it is necessary to consider multi-factor problems. However, before introducing optimization for handling multiple factors, we should study Chairman Mao's conclusion: "In any complex process involving numerous contradictions, there must be one that is principal, playing a leading and decisive role, while others are secondary and subordinate."

While optimization is indeed more suitable than traditional exhaustive (or permutation methods) for dealing with multi-factor problems, it must be noted that as the number of factors increases, the number of experiments also increases rapidly (though not as fast as with traditional methods). Therefore, to speed up the process, save manpower and resources, and reduce the number of experiments, identifying the principal contradiction becomes the key of keys. At the very least, we should temporarily set aside less influential factors and concentrate efforts on a few essential and decisive ones for in-depth research.

For example, after quenching a certain metal alloy component, a layer of oxide scale forms. We want to remove the oxide scale without damaging the smoothness of the metal surface. One method is called acid pickling, where several acids are mixed into a solution, and the metal component is soaked in it. The goal is to remove the oxide quickly without compromising surface smoothness.

We won't go into which acids to choose. Let's say we know that nitric acid and hydrofluoric acid must be used. The question is that what is the best formula? Specifically, how should 500 mL of pickling solution be composed?

Consider how many factors are involved: How much nitric acid? How much hydrofluoric acid? How much water? What temperature? How long? Should it be stirred? If so, at what speed and for how long? That gives seven factors. If each factor is divided into ten levels, an exhaustive method would require 10^7 tests—that's ten million experiments! Even if optimization only requires one ten-thousandth of that, that is 1,000 tests—still far too many.

Let's see how the experimenters, following Chairman Mao's directive to identify the principal contradiction, analyzed the problem.

Out of the total 500 mL, once the amounts of the two acids are set, the amount of water is automatically determined—so water is not an independent factor.

Next, since the solution is used immediately after mixing, the temperature doesn't vary much and can be disregarded.

As for time, if it refers to the time between preparation and use, it's irrelevant because the solution is used right away. If it refers to the pickling time, that's not a factor but an indicator—in this case, it only takes three minutes, so it's not a concern either.

Finally, whether or not to stir can be temporarily ignored.

In the end, only two factors remain: the amount of nitric acid and the amount of hydrofluoric acid. So, with just 14 experiments in one day, they solved the problem. Otherwise, it could have taken months or even years.

One additional point of clarification: Sometimes, a formula involving three components might be misunderstood as involving three factors. In fact, there are only two independent factors (variables).

Also, the duration of pickling is not a factor but an indicator—it's a dependent variable, not an independent one.

Those using optimization must pay attention: When analyzing the problem, it's critical to determine which variables are truly independent. Experience shows this is where mistakes happen. And once the independent variables are identified, it's equally important to assess which ones are primary.

▸ Two-Factor Problems

Suppose there are two factors to consider: One is a content range from 1,000 grams to 2,000 grams. The other is a temperature range from 5,000°C to 6,000°C.

The method we use is this: First, fold the paper vertically—say, at 1,500 grams. Then, with 1,500 grams fixed, use the single-factor method (as described in "Single-Factor Optimization") to find the optimal temperature. Suppose this optimal point is marked "×." Next, fold the paper horizontally—say, at 5,500°C—and, with that temperature fixed, use the same single-factor method (as described in "Single-Factor Optimization") to find the optimal content. Suppose this point is marked "○" (fig. 8).

Compare the results at "○" and "×." Whichever gives the better result, keep that half of the paper and discard the other. (If the result at "×" is better, cut off the lower half; if "○" is better, cut off the left half.) On the remaining half-sheet, repeat the process.

Of course, the more factors involved, the more complex the problem becomes. But complexity also offers room for flexible thinking. For example, once we find the "×" point, we may abandon the folding method and instead draw a horizontal line through "×," and perform new trials along that line using the single-factor method (as described in "Single-Factor Optimization") to find the optimal point "▭." Then, draw a vertical line through "▭" and continue testing along that line, and so on (fig. 9).

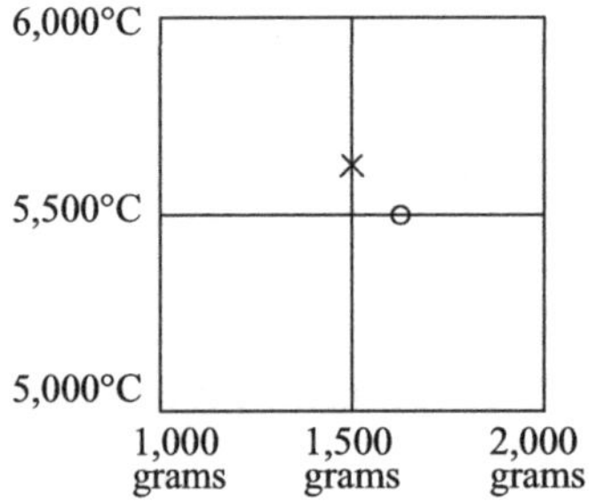

Figure 8

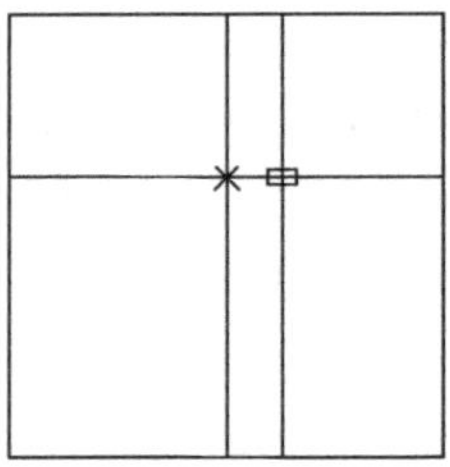

Figure 9

For example, a factory once dealt with a problem just like the one discussed here—using pickling solution to remove oxide scale from metal components. After analysis, the problem was redefined as: How should 500 mL of pickling solution be composed (i.e., how much water, nitric acid, and hydrofluoric acid) to achieve the best effect?

Based on experience and relevant data, they initially set the ranges as nitric acid: 0 to 250 mL and hydrofluoric acid: 0 to 25 mL. The rest is water. This is a two-factor problem.

If the tests were conducted using a permutation method. Nitric acid: divided into 50 intervals of 5 mL. Hydrofluoric acid: divided into 13 intervals of 2 mL. That would require 50 × 13 = 650 experiments—a huge cost in time and resources. However, using optimization, the final best hydrofluoric acid amount turned out to be 33 mL, which exceeded the originally defined range. So even if they had done all 650 experiments, they wouldn't have found this optimal formula.

The steps for guiding an experiment applying optimization are as follows:

Step one: Fix hydrofluoric acid in the middle of its range (0–25 mL), say at 13 mL. Then apply the "Single-Factor Optimization" single-factor method to optimize nitric acid. Mark 0–250 mL on a paper strip. Starting from 0, use the 0.618 ratio to find point A (155 mL), and test it. Fold the strip so that point A is mirrored to point B (95 mL) and test it (fig. 10). Compare A and B—if A is better, discard the left side (B's side); the new range is 95–250 mL. Repeat: fold, find point C (190 mL), and test and compare with A (fig. 11). If A is still better, discard right side (above 190 mL), narrowing to 95–190 mL. Continue folding and testing (fig. 12). After the 5th test, the optimal nitric acid amount is found to be 165 mL.

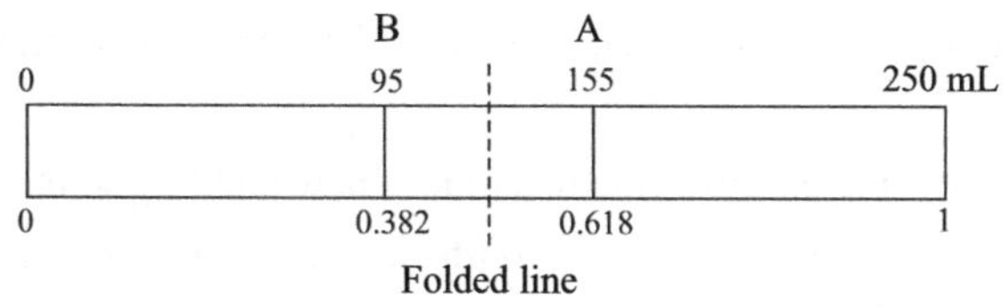

Figure 10

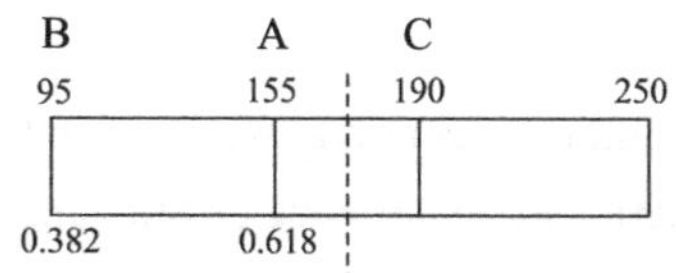

Figure 11

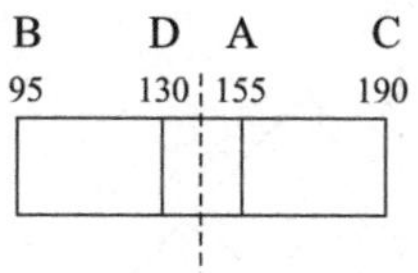

Figure 12

Step two: Fix nitric acid at 165 mL, and apply the same method to optimize hydrofluoric acid. It turns out that at the boundary point of 25 mL, the quality is already better, indicating the original range may have been too limited. So, they expand the range to 25–50 mL and continue optimization. By the 9th test, the optimal hydrofluoric acid amount is found to be 33 mL. In total, only 14 experiments were needed to find a formula that meets production needs. The experiment could end here, but if desired, nitric acid can be re-optimized with hydrofluoric acid fixed at 33 mL—and so on—until the truly optimal formula is found.

This example shows that optimization not only finds the best solution efficiently and effectively, but it also helps to correct errors in the initial assumed ranges based on experience.

Note 1: Regarding the alloy pickling solution formulation problem mentioned above: Over the past two years, two rounds of experiments were conducted. In 1968, the experiment failed. In 1969, after numerous trials, a barely usable pickling solution was finally found; however, it required half an hour of pickling time and brushing.

In contrast, by using optimization this time, a new pickling solution formula was found in less than one day through only fourteen experiments. When alloy materials were immersed in this new solution, the reaction occurred immediately, and after just three minutes, the oxide layer naturally peeled off, leaving the material surface smooth and without any signs of corrosion.

Note 2: Let x represent the amount of nitric acid, and y the amount of hydrofluoric acid.

Based on experience and relevant data, we assume $0 \leq x \leq 250$ (mL) and $0 \leq y \leq 25$ (mL). But what if there's no experience or relevant data, and we're only given the following constraint (fig. 13): $x + y \leq 500$, $0 \leq x$, $0 \leq y$.

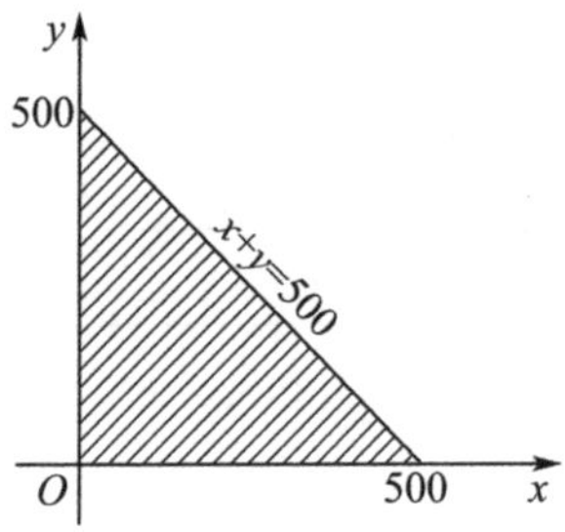

Figure 13

How do we proceed in selecting and combining the components under this condition? In this case, the two-factor method described above is still applicable. However, we must ensure that points outside the triangle defined by the constraint are not considered. A better approach is to change the variables.

Let $z = x + y$, $x = tz$. In other words, let z ($0 \leq z \leq 500$) represent the total amount of acid added, and let t ($0 \leq t \leq 1$) represent the proportion of nitric acid in the total acid, treated as the independent variable. This transforms the problem into one of finding the optimal solution within the rectangle defined by $0 \leq z \leq 500$, $0 \leq t \leq 1$.

▸ Multi-Factor Problems

(This section can be skipped at first glance. After gaining some practical experience and fully mastering the method for handling one or two factors, come back and try applying the ideas here.)

Some may ask that since the "folded paper method" relies on the paper having only length and width, it's only suitable for dealing with two-factor problems—so what about problems with more than two factors? If you've studied math, you might use the term "dimensionality reduction" to approach this. Once you understand how to reduce dimensions, the problem becomes manageable. What we've covered so far about two-factor methods is essentially a way of reducing a two-dimensional problem into one dimension.

Up to now, we've been folding rectangles. Now we abstract that into folding a rectangular prism—in other words, we split a cuboid into halves. As you know, there are three ways to bisect a cuboid. On each of these three planes of bisection, we search for the optimal point. Each of these is essentially a two-factor optimization problem (with one factor held fixed).

From these three planes, we identify three optimal points. We then compare these points to determine which one yields the best result, and we retain the quarter of the original cuboid that contains this point, continuing the process using the same method.

For example, as shown in figure 14, let's say we want to find the optimal point within the cube: $0 \le x \le 1, 0 \le y \le 1, 0 \le z \le 1$.

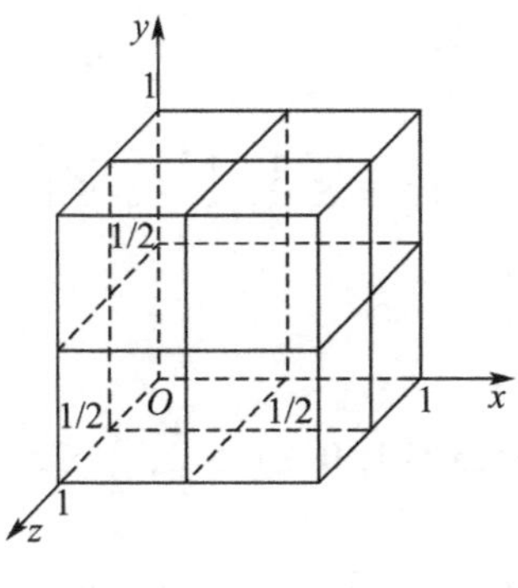

Figure 14

On the three planes,

$$x=\frac{1}{2},\ 0\le y\le 1,\ 0\le z\le 1$$

$$0\le x\le 1,\ y=\frac{1}{2},\ 0\le z\le 1$$

$$0\le x\le 1,\ 0\le y\le 1,\ z=\frac{1}{2}$$

we use the two-factor method on each plane to find the best points:

$$\left(\frac{1}{2},y_1,z_1\right),\ \left(x_2,\frac{1}{2},z_2\right),\ \left(x_3,y_3,\frac{1}{2}\right)$$

Whichever point among the three is best, we keep the quarter volume that contains it. For instance, if

$$\left(\frac{1}{2},y_1,z_1\right)$$

is best, and

$$0\le y_1\le\frac{1}{2},\ 0\le z_1\le\frac{1}{2}$$

then we continue the search in

$$0\le x\le 1,\ 0\le y\le\frac{1}{2},\ 0\le z\le\frac{1}{2}$$

If

$$0\le y_1\le\frac{1}{2},\ \frac{1}{2}\le z_1\le 1$$

we continue in

$$0\le x\le 1,\ 0\le y\le\frac{1}{2},\ \frac{1}{2}\le z\le 1$$

and so on. The retained volume is always one-fourth of the previous.

In actual operations, once optimal points are identified on two of the three planes, you can first eliminate half of the volume, then process the remaining plane.

Special Case Problems

Cases Where Multiple Experiments Can Be Done at Once

For example, suppose you can run four experiments at the same time. What should you do? Based on this condition, we suggest the following method:

1. Divide the interval into five equal parts (fig. 15) and perform tests at the four internal division points.

Figure 15

2. Compare the four results. Retain the best-performing point and its neighboring two segments. Then divide the retained section into six equal parts (fig. 16) and conduct another experiment at the marked point "×."

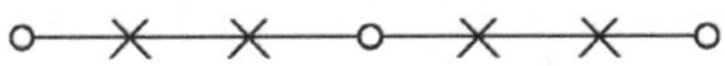

Figure 16

3. Continue this way—each time, retain the best point and its adjacent intervals—and keep repeating. This allows you to gradually zero in on the optimal point.

This method was actually suggested by an experienced master worker at a factory. In fact, it can be proven to be optimal. However, note that this approach works best when the number of tests per batch is even. For odd-numbered test batches, things become a bit more complicated (for example, using the 0.618 rule for single tests), which we won't go into here.

Some references claim that the "optimization" only applies when tests are done one at a time, and that if multiple tests must be done in parallel, one should fall back on traditional "experimental design" approaches. This opinion is debatable and worth discussion.

▸ Equal Division Method

In practice, you might face this kind of problem: A certain product relies on a precious metal. We know that using 16% of the metal yields acceptable quality. The question is that can we reduce it? Maybe even further? The goal is to maintain product quality while lowering costs.

We recommend adopting the equal division method instead of the 0.618 method. First, test at the midpoint, say 8% (fig. 17). If 8% still meets quality standards, eliminate the right half. If not, eliminate the left half. Then test at 4% (midpoint of the new interval). If it fails, eliminate the right half. Next test at 6%. If it passes, test at 5%. If 5% also passes, you can settle on 6%, leaving a safety margin.

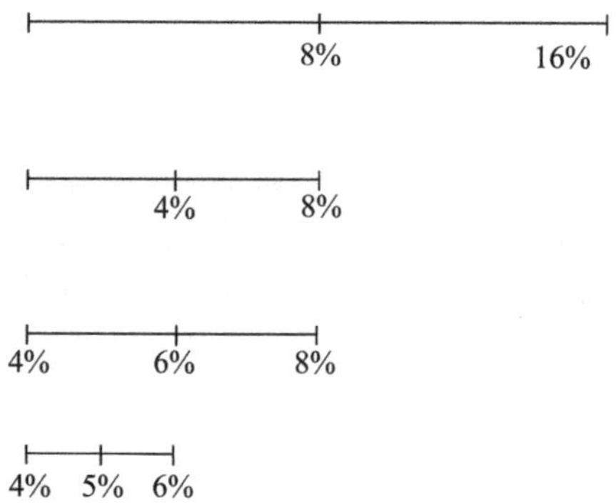

Figure 17

This method has already been widely applied in many factories.

▸ Parallel Lines Method

In this case, we are dealing with two factors: one is temperature, the other is time. Though furnace temperature is difficult to adjust, time is easy to control, we use the parallel lines method (fig. 18).

We first fix the temperature at 0.618 of its range, and then find the optimal point for different times—this point is marked "∘." Next, we adjust the temperature to 0.382, fix it again, and test different times to find another optimal point, marked "×." After comparing the two, if the "∘" point is better than "×," we eliminate the lower part of the range. Then we use the folding method to determine the next temperature value, and so on.

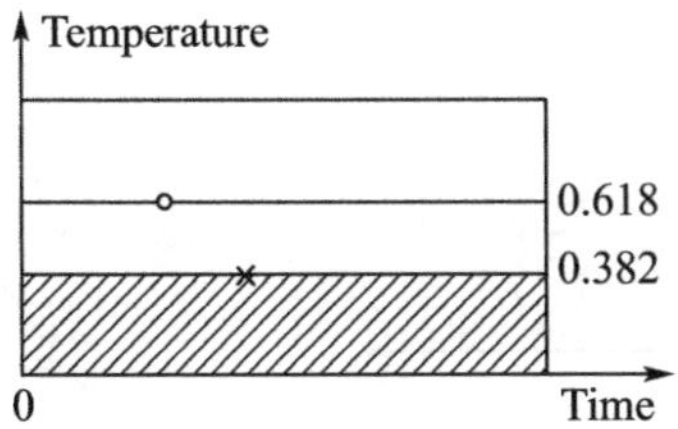

Figure 18

This method was an invention developed in a factory based on real operational conditions.

Gradient (Steepness) Method

Suppose you do an experiment at point A and get a result of a, and then at point B you get b. If $a > b$, then

$$\frac{(a-b)}{\text{distance between A and B}}$$

This value is called the steepness of ascent from B to A (fig. 19).

In one chemical factory, we encountered a problem like the one shown in figure 20. It involved two factors. We first performed two tests along the horizontal line (① and ②), then shifted to the vertical line and conducted two more tests (③ and ④). We found that point ④ was excellent, and point ② was very poor. At this point, we stopped testing along the vertical and horizontal lines. Instead, we tested a new point ⑤ on the line connecting ② and ④, and got even better results—exceeding expectations.

In summary, this is a steepness problem. We can compute the steepness from points ① to ④, ② to ④, and ③ to ④. The steepest path tells us the best direction to move in—just like climbing uphill.

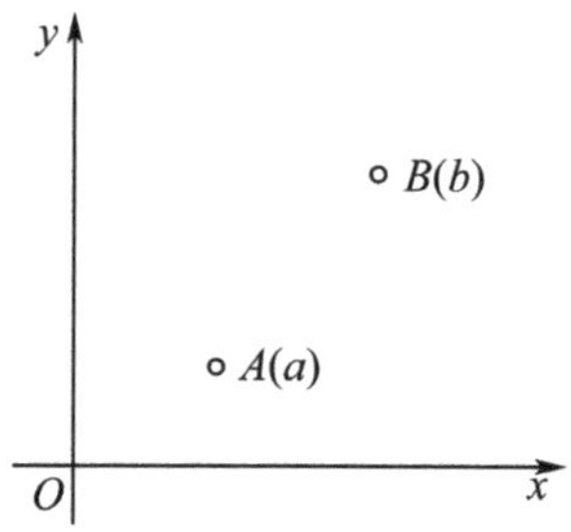

Figure 19

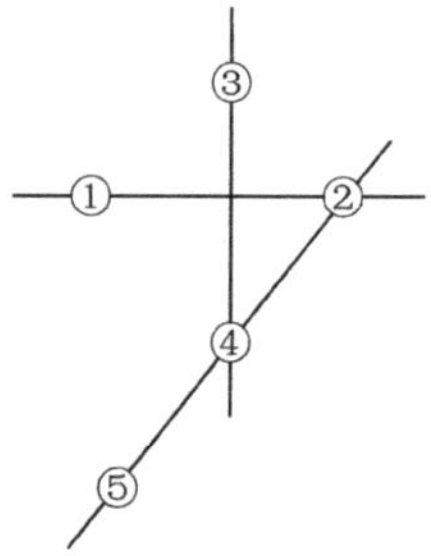

Figure 20

This method has been used in a factory, where steep directions were identified based on available experimental data—and testing in those directions led to optimal solutions. By experimenting in this direction, we found the most satisfactory point.

▸ Blind Man Climbing a Mountain Method

Imagine a blind person on a mountain who wants to reach the summit. What should he do? From his current location, he taps forward with a cane. If it's higher, he takes a step forward. If not, he tries left. If that's higher, he steps left. If not, he tries behind. If that's higher, he steps back. If still no elevation, he tries right. If that's higher, he goes right. If none of the directions are higher, he stays put. In short, if it's higher, step toward it. Step by step, the blind person will eventually reach the top.

This method is useful when large adjustments aren't possible. Of course, in practice, we don't need to test in just four directions (north, south, east, and west). For example, after stepping north once and then east once, we might collect data points z_0, z_1, and z_2 (fig. 21). The steepness from z_1 to z_2 is $z_2 - z_1$, and from z_0 to z_2 is

$$\frac{z_2 - z_0}{\sqrt{2}}$$

If

$$\frac{z_2 - z_0}{\sqrt{2}} > z_1 - z_0$$

then why not try moving directly in the direction of vector $\overrightarrow{z_0 z_2}$ (fig. 22)? The more points we gather, the more we can infer which direction is uphill.

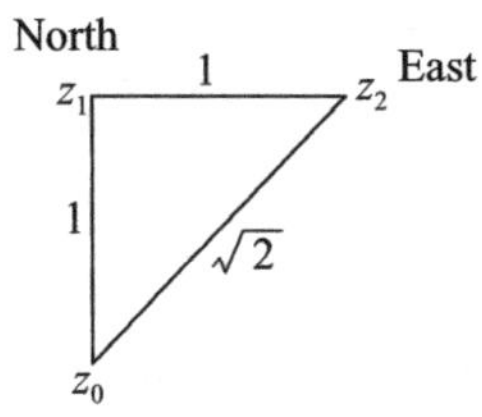

Figure 21

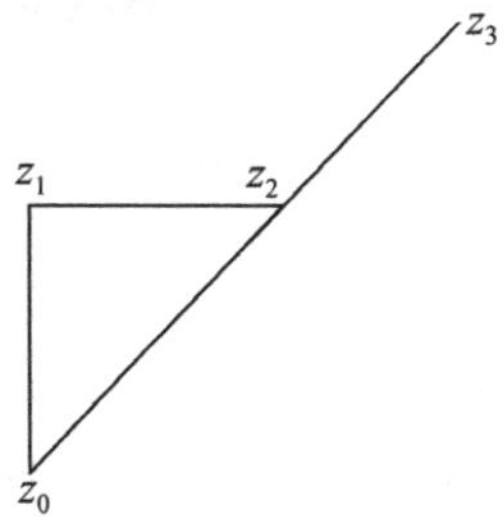

Figure 22

This method is suitable in situations where a system is already in production and large adjustments are not practical.

▸ What If the Function Is Not Unimodal?

Some may say that the methods described above only apply to unimodal cases—i.e., when there's only one peak and all neighboring points are worse. But what if the situation is multimodal—where multiple peaks exist? We recommend as follows:

1. Try the unimodal method anyway. Start by assuming the function is unimodal and try to find one "peak." If that peak meets your production needs, begin production. Later, when time permits, search for other possibly higher peaks—this is called zonal or regional searching.

2. Use sparse, evenly distributed trials. Start by doing a batch of sparse and well-distributed tests to see if multiple peaks exist. If you detect a multimodal pattern, then conduct zonal optimization. For single-variable problems, it's best to divide intervals according to the golden ratio: $\alpha:\beta = 0.618:0.382$.

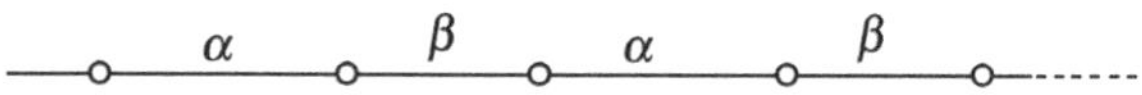

Figure 23

For example, with three division points, you could choose them as in figure 24.

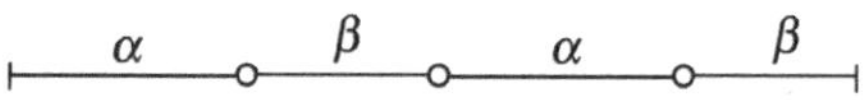

Figure 24

The remaining segment then takes the form shown in figure 25, which is ideal for applying the 0.618 method.

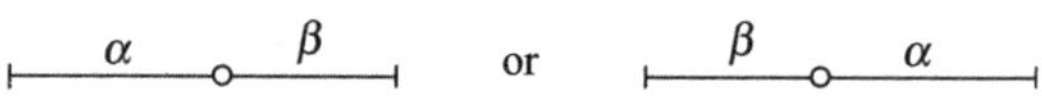

Figure 25

Don't worry that our method is worse than the brute-force method (also known as permutation or combinatorial enumeration). At most, after adopting optimization, you might need to supplement with a few tests that the brute-force method would have required anyway.

In real-world research—especially in exploratory scientific projects—you might encounter very complex problems. For example, saddle points: a saddle-shaped point, which is a maximum in the left-right direction, but a minimum in the front-back direction. If you follow conventional methods, you might never finish the tests. But with optimization, such problems have been solved in a week or two. This has been observed in numerous chemical engineering cases.

■ Supplement

We briefly explained general methods in the first part using plain language, and listed some specific methods in the second part. If these two parts are still not sufficient during application, this third part may serve as a reference. If you can't fully understand everything right away, don't worry—just use what you can. Through continuous practice and reflection, you will make progress. As for reading complete theoretical books, it's best done after gaining some practical experience.

▸ This Is a Problem of Finding a Maximum (or Minimum) Value

For those who have studied mathematics, this is a problem of finding the maximum (or minimum) of a function. For example, a certain quality index T depends on three factors, that is $T = f(x,y,z)$. The main problem is to find the maximum value of $f(x,y,z)$ within the range: $a \leq x \leq p, b \leq y \leq q, c \leq z \leq r$. Some may think this is a familiar topic from calculus textbooks. But in reality, the problem lies in feasibility. First of all, you need to know the exact expression of $f(x,y,z)$. Even if you know the analytical form, you must solve the system:

$$\frac{\partial f}{\partial x} = 0, \frac{\partial f}{\partial y} = 0, \frac{\partial f}{\partial z} = 0$$

This may involve transcendental equations that are not easy to solve; and even if you do solve them, you still have to judge whether the solution is indeed the maximum within the whole region.

However, simple expressions for $f(x,y,z)$ are rare and may not even be discovered or expressible. For example, in the plain language section above, how is the percentage of people who say "it tastes good" a function of the amount of baking soda added?

Some might suggest using statistical regression to derive a formula and then finding the maximum value. But statistics always require extensive experiments, and the calculations are no piece of cake. Moreover, regression functions often lose accuracy (usually assuming first- or second-degree polynomials). If we're planning to do a large number of experiments anyway, why not directly adopt optimization? Besides, it can greatly reduce the number of required experiments!

▸ The Origin of 0.618

0.618 is a three-decimal approximation of

$$W = \frac{-1+\sqrt{5}}{2}$$

Depending on practical needs, you can use 0.6, 0.62, or a more precise value of 0.618.

This number W has a special property: $1 - W = W^2$ (The solution to this equation is indeed $\frac{-1+\sqrt{5}}{2}$).

W and $1 - W$ divide the interval [0,1] as shown in the figure below:

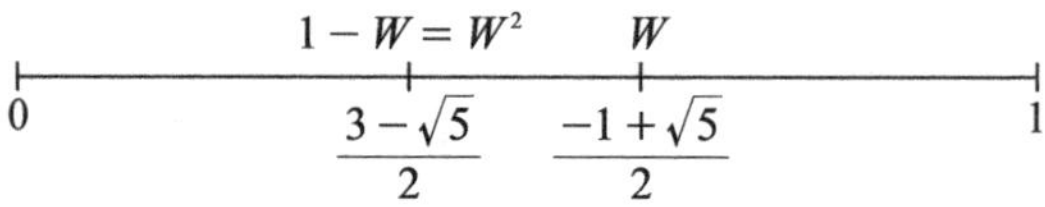

Figure 26

No matter which segment you discard (either $[0,(1-W)]$ or $[W,1]$), the remaining segment still contains a point that has the same proportional position as one of the original points ($1 - W$ or W) in the interval [0,1]. Specifically, in the case where $0 < 1 - W < W < 1$. If you discard the right segment $[W,1]$, the remaining is $0 < 1 - W = W^2 < W$. This is simply a scaled version of [0,1] by a factor of W.

Similarly, if you discard the left segment $[0, (1 - W)]$, the remaining segment is

$$1 - W < W = (1 - W) + W(1 - W) < 1$$

The total length of this segment is still W, and the distance from W to 1 is W times $(1 - W)$.

This method is the golden section method from plane geometry. Hence optimization is also known as the golden section method. During the Middle Ages in Europe, there were tales that windows made according to the golden section ratio were the most aesthetically pleasing (i.e., in the ratio 0.382:0.618).

▸ "Back-and-Forth Adjustment Method"

Readers may think the previous section already explained the origin of

$$W = \frac{-1+\sqrt{5}}{2}$$

but the more accurate question is that in the interval $[a,b]$, if there is a unimodal function $f(x)$, how can we find its peak (without knowing the actual expression of $f(x)$)?

First, choose a point x_1 to test, obtaining $y_1 = f(x_1)$, then choose a second point x_2 to test, getting $y_2 = f(x_2)$. If $y_2 > y_1$, discard $[a,x_1]$; if $y_1 < y_2$, discard $[x_2,b]$. In the remaining interval, choose a new point x_3 (possibly between x_1 and x_2), test $y_3 = f(x_3)$, If $y_3 < y_2$, discard $[x_3,b]$; in the remaining interval (x_1, x_3), choose another point x_4 (as shown in fig. 27) … Continue this process.

No matter how randomly you proceed, you'll eventually find the maximum of $f(x)$. But the question is that how should we choose x_1, x_2, … to achieve the fastest results (i.e., minimize the number of tests for any $f(x)$)? Answering this requires some not-too-advanced mathematical knowledge (for example, chapter 1 of *Introduction to Advanced Mathematics*), which we won't elaborate here.[1] But it must be noted that the so-called proofs in some foreign literature are not actually rigorous.

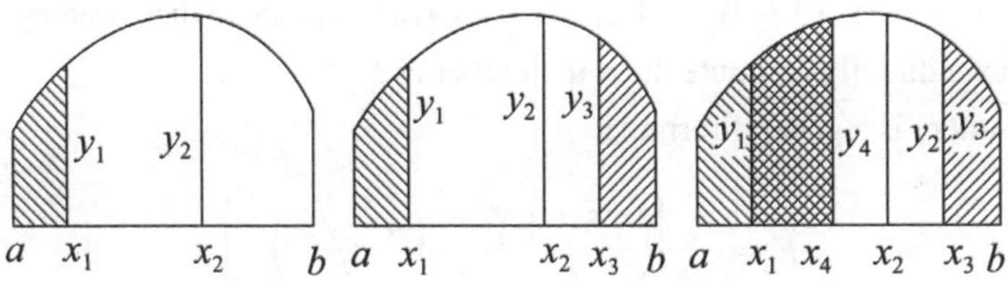

Figure 27

Fraction Method

In the history of Chinese mathematics, there was a brilliant chapter concerning the value of π. The great mathematician Zu Chongzhi (429–500) made two important contributions: He expressed the value of π using decimals: $3.1415926 < \pi < 3.1415927$. 2. He used the fraction $\frac{355}{113}$ to represent π. This fraction is accurate to six decimal places, and among all fractions with denominators less than 33,102, there is none closer to π than this one.

1 For those familiar with mathematical induction, it is recommended to first use it to prove the expression for F_{n+1} and that the fraction method is the best. Then, go on to prove that the limit of the fraction method is the golden section. However, the drawback of induction is that you need to know the conclusion in advance.

Such a fraction is called a best approximating fraction (you may refer to *On the Topic of Zu Chongzhi's π*).

We now deal with

$$\frac{\sqrt{5}-1}{2}$$

using two methods. One is the decimal method: 0.618. The other is the fraction method, as mentioned in a certain book, which gives continued approximations of the number:

$$\frac{3}{5}, \frac{5}{8}, \frac{8}{13}, \frac{13}{21}, \frac{21}{34}, \frac{34}{55}, \frac{55}{89}, \frac{89}{144}, \cdots$$

These fractions are derived from the sequence: 1, 1, 2, 3, 5, 8, 13, 21, 34, 55, 89, 144, ... This sequence follows the rule: $1+1=2$, $1+2=3$, $2+3=5$, $3+5=8$, $5+8=13$, $8+13=21$, $13+21=34$, ... Do we have to calculate them one by one like this? Can we directly compute the n-th number F_n?

Yes—there is a general formula:

$$F_n = \frac{1}{\sqrt{5}}\left[\left(\frac{\sqrt{5}+1}{2}\right)^{n+1} - \left(\frac{1-\sqrt{5}}{2}\right)^{n+1}\right]$$

(Readers may refer to *On the Topic of Pascal's Triangle* for this formula. With it, one can also use mathematical induction for proof.) It's also easy to calculate

$$\lim_{n\to\infty}\frac{F_n}{F_{n+1}} = W = \frac{\sqrt{5}-1}{2}$$

Because of this asymptotic property, readers can see that the fraction method and the golden section method yield similar results. Only in very specific cases would the fraction method save one experiment.

If the number of trials is strictly limited, we can use a fraction instead of 0.618. For example, if ten experiments are allowed, we suggest using $\frac{89}{144}$; for 9 trials, use $\frac{55}{89}$, and so on. This approach is suitable only when the cost of a single experiment is exceedingly high.

Parabola Method

For technical precision, neither the golden section method nor the fraction method takes into account the actual values of experiment results—they only compare relative magnitudes. Can we make use of those trial values?

For instance, after obtaining three experiment results, we could fit a parabola through those three points and use the vertex of that parabola as the basis for our next experiment. Specifically, if we perform tests at three points x_1, x_2, x_3 and obtain data y_1, y_2, y_3 (fig. 28), we can use the interpolation formula:

$$y = y_1 \frac{(x-x_2)(x-x_3)}{(x_1-x_2)(x_1-x_3)} + y_2 \frac{(x-x_1)(x-x_3)}{(x_2-x_1)(x_2-x_3)} + y_3 \frac{(x-x_1)(x-x_2)}{(x_3-x_1)(x_3-x_2)}$$

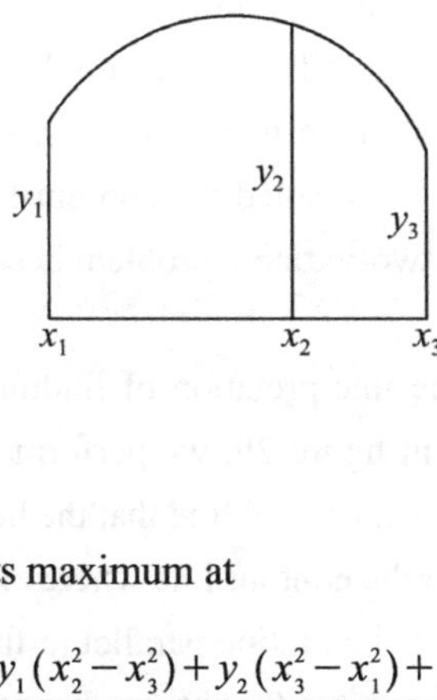

Figure 28

This function reaches its maximum at

$$x_0 = \frac{1}{2} \cdot \frac{y_1(x_2^2-x_3^2)+y_2(x_3^2-x_1^2)+y_3(x_1^2-x_2^2)}{y_1(x_2-x_3)+y_2(x_3-x_1)+y_3(x_1-x_2)}$$

So for the next trial, we take $x = x_0$. (It is best to do this when y_2 is greater than both y_1 and y_3.) If $x_0 = x_2$, then the method must be adjusted. For example, take

$$x_0 = \frac{1}{2}(x_1 + x_2)$$

Two Variables and Contour Lines

As the number of variables increases, the problem becomes more complex and more difficult. But the more complex the problem, the more it challenges our thinking—and the more room there is to apply our skills. As mentioned in section 2, we don't necessarily need to finish testing one diagonal line before moving to another. Instead, we can perform one or two trials on each line and then utilize the concept of steepness.

Some have also suggested that instead of performing the first round of trials along the center line (the midpoint), we could use the 0.618 line and apply the single-factor method to find the optimal point along that line. This is a good suggestion—it could save one trial in the next round. We also mentioned earlier that in situations where temperature is hard to adjust but time is easy to control, the parallel-line method may be better. These flexible approaches show that when the problem is complex, flexible thinking becomes even more essential.

Let's still start by discussing the case with two variables.

Assume we conduct trials within a unit square: $0 \leq x \leq 1$, $0 \leq y \leq 1$ and seek the maximum of $f(x,y)$. Geometrically, we can think of $f(x,y)$ as the height at the point (x,y). If we draw curves where $f(x,y)$ takes the same value, these are called contour lines. A curve $f(x,y) = a$ is called the contour line at elevation a. Thus, the geometric representation of a two-variable problem becomes a contour map, similar to a topographic map.

Let's revisit the geometric interpretation of finding the optimal point along a line. For example, as shown in figure 29, we perform trials along the line (1) $x = 0.618$ using the single-factor method, and find that the best point lies at point A, with a value of a. This point lies on the contour line where elevation = a. Next, along the horizontal line through point A (i.e., a line parallel to the x-axis), we search for the best point again. Suppose this point is B, with a value of b, where $b > a$. This means point B lies on a higher contour line $f(x,y) = b$. Then, we continue by moving along the vertical line through B (fig. 30) to find the next optimal point … This method is essentially a step-by-step approach where we move from one contour line to a higher one, until we reach the highest point—the peak.

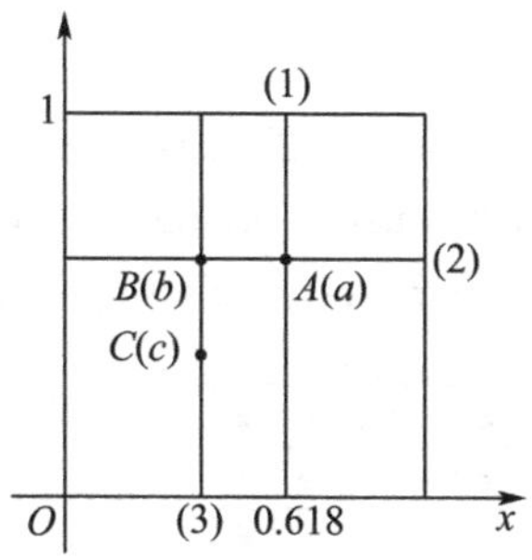

Figure 29

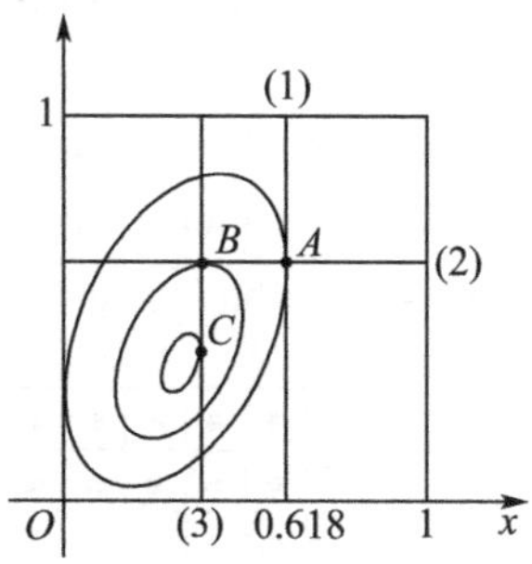

Figure 30

Note that some people think that if a point is optimal in both the horizontal and vertical directions, it's a "dead point" (i.e., a point where the method can no longer continue). But this is a misunderstanding. This is not a "dead point"—on the contrary, it is the most meaningful point. (You can verify this by checking whether

$$\frac{\partial f}{\partial x} = 0, \frac{\partial f}{\partial y} = 0$$

that tells you the significance of this point.)

Once we have this geometric model, it can inspire many methods. One such method discussed earlier in section 2 is steepness. Other methods include steepest ascent method (gradient method), block cutting method, parallel tangent line method, etc.

There are many methods for multivariable optimization, which we won't elaborate on here. However, it must be noted that in capitalist countries, many optimization methods have fancy names but essentially do the same thing—designed for the sake of proprietary terms and patents. This is a natural phenomenon under the capitalist system. But we must investigate beyond the names, analyze their true nature, and decide what to adopt or discard.

▸ Statistical Experimental Method

Take a square (or rectangle) and divide each side into 100 parts, resulting in 10,000 small squares. Take the center point of each square—giving us a total of 10,000 points. Our objective is to find the point where the test result is optimal (fig. 31).

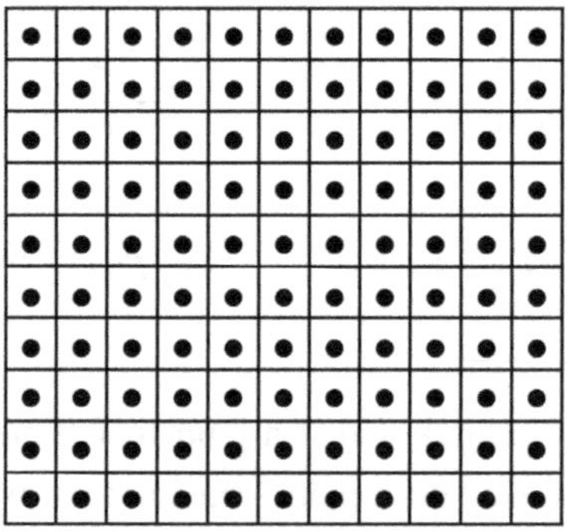

Figure 31

If we consider an easier problem—finding a point better than 8,000 of the other points—we suggest the following method:

Label these 10,000 points from 1 to 10,000. Then, prepare a draw-bag with 10,000 number slips. Whichever number is drawn, conduct an experiment at the corresponding point. This method is known as the Statistical Experimental Method, also referred to in Western literature as the Monte Carlo method.

The principle behind it is that suppose a bag contains 2,000 white balls and 8,000 black balls. The probability of drawing a white ball is 2,000 / 10,000 = 0.2 = 20%. The probability of drawing a black ball is $1-0.2=0.8$. Probability of drawing two black balls in a row is $(0.8)^2=0.64$. Probability of drawing four black balls in a row is $(0.8)^4=(0.64)^2=0.41$. Probability of drawing eight black balls in a row is $(0.8)^8=(0.41)^2=0.17$. Probability of drawing ten black balls in a row is $(0.8)^{10}=(0.8)^8(0.8)^2=0.11$. Thus, the probability of drawing at least one white ball in ten tries is $1-0.11=0.89$.

In other words, you have an 89% chance of finding a point better than 8,000 others by randomly conducting ten trials.

Advantages of this method are as follows: It works regardless of the shape of the function—whether it's jagged, smooth, or highly irregular. It's largely unaffected by the number of variables involved. However, it is, after all, a statistical method, which means there's always a degree of luck involved. To rely on the law of large numbers, many trials are usually needed.

There is also the issue of how to draw numbers randomly. Besides using a physical draw-bag, there are also so-called random number generators: One kind uses a Geiger counter to detect particle counts, distinguishing odd and even numbers to produce binary digits. Another uses noise amplifiers. These machines can generate random numbers rapidly, but for experimental purposes, such high speed is not really necessary.

A better approach is the number theory method (see *Numerical Integration and Its Application* by Hua Luogeng and Wang Yuan [Science Press, 1963]). This method requires no equipment. Its error is deterministic, not probabilistic, unlike the two machines above.

That said, readers must analyze and understand this method before applying it—don't use it lightly.

► Estimating Effectiveness

If we divide the interval [0,1] into $n + 1$ equal parts and conduct n trials, we can determine that the optimal point lies within an interval of length $\frac{2}{n+1}$. If we want an estimated accuracy of δ, then we need a number of trials n such that

$$\frac{2}{n+1} < \delta$$

This means n is on the order of $\frac{1}{\delta}$.

For the golden section method, after n trials, the optimal point will be within an interval of length $(0.618)^{n-1}$. If we want this to be less than δ, it's easy to calculate

$$n > 4.8 \log \frac{1}{\delta}$$

So here, the number of trials is on the order of

$$\log \frac{1}{\delta}$$

For k variables, the equal division method has complexity on the order of

$$\left(\frac{1}{\delta}\right)^k$$

The multi-variable golden section method has complexity on the order of

$$\left(\log\frac{1}{\delta}\right)^k$$

We also have methods that achieve complexity on the order of

$$k\log\frac{1}{\delta}$$

In fact, there are methods with complexity as low as

$$\left(\log\log\frac{1}{\delta}\right)^k$$

But since having k in the exponent is undesirable, an even more efficient method achieves complexity

$$\frac{k^2}{\log k}\log\log\delta$$

However, be cautious that the method that is theoretically the most precise is not always the most practical in real-world applications. The most important thing is to select a method that is simple, fast, and suitable to the specific problem.

Note the estimated precision δ beforehand is not always reliable. Sometimes, if the function is flat, even a large interval can't distinguish between better or worse results. Other times, if the function is steep, even a small interval may make a big difference. In other words, we are dividing the domain x, but what we're really trying to distinguish is the function value $y = f(x)$, which we don't yet know. Thus, this estimation can only serve as a reference.

As for the "fraction method," its advantage is that if the number of trials happens to match exactly with one of the values in the Fibonacci sequence F_n, then you can save one trial compared to the golden section method. However, if suitable data happens to come earlier, there is no saving. And if it doesn't, and you need to go further, you may end up doing one or two more trials instead.

A Shining Example for Intellectuals*: In Commemoration of the 80th Anniversary of Martyr Wen Yiduo's Birth

This year marks the 80th anniversary of the birth of Wen Yiduo, a renowned Chinese poet, scholar, and professor. Though it has been over three decades since he was martyred, his radiant image—slamming the table in defiance, furrowing his brows in anger, facing the Kuomintang's guns, choosing to fall rather than submit—still lives vividly in our hearts. Wen sacrificed his life for the founding of the People's Republic of China, and the Chinese people will forever remember him.

Wen was both a respected teacher and a dear friend to me. From the Second Sino-Japanese War to the Chinese Civil War, we both lived and worked at the National Southwest Associated University, which was then known as a stronghold of democracy. That was an extraordinary period, and the lessons I learned from my many interactions with Wen during that time remain unforgettable to this day.

Slamming the Table in Righteous Rage, Denouncing the Reactionaries

Dark clouds hang low over the still waters,
The red candle's light pierces the stars above.
On the road from Nanjing, we hear the tragic news,

* First published in *Wen Yiduo Memorial Collection*, Sanlian edition, 1979.

The devil's hand dares to take Wen Yiduo!

This short poem is but a drop in the ocean of writings that emerged following the shocking "Li-Wen Incident" of July 1946.

I heard of Wen's martyrdom on the train from Nanjing to Shanghai. The Kuomintang reactionaries were so cruel and unrestrained that I was left speechless. Wen's death for the nation filled me with profound grief.

I left Kunming at the end of June 1946. At that time, the Kuomintang reactionaries were preparing to tear up the still-wet ceasefire agreement and launch a full-scale civil war. Simultaneously, they were deploying all kinds of despicable and shameless tactics to suppress the patriotic democratic movement. Kunming was already facing a "storm is brewing" atmosphere—agents swarmed with murderous intent, ready to strike at any moment. The political climate was highly volatile and people were uneasy. With the announcement of the National Southwest Associated University's closure, many began to scatter. The reactionaries saw this as a chance to dismantle this democratic bastion and crush the surging democratic movement. Yet, amid this blood-soaked storm, Professors Li Gongpu and Wen Yiduo stood tall and unyielding. With furrowed brows and fierce defiance, they became thorns in the eyes of the reactionaries. Spies openly declared a bounty of 400,000 yuan for Wen's head.

Before I left Kunming, I had just returned from a visit to the Soviet Union. I was deeply concerned for his safety. After giving my report on the trip, I advised him: "The situation is really tense. Everyone is leaving. You need to be extra cautious." Wen calmly replied: "If we are to fight, someone will have to fall. When one man falls, thousands will rise. The more tense the situation, the more I must shoulder this responsibility. 'If the people do not fear death, how can you threaten them with it?' Are we, today's intellectuals, truly lesser than those of ancient times?"

Never could I have imagined that just as I set foot out of Kunming, Wen Yiduo, following in the footsteps of Li Gongpu, would be murdered by the reactionary forces.

"When one man falls, thousands will rise!" Wen Yiduo, these were your words, and you truly lived by them. On July 11, when Li Gongpu was assassinated, you scornfully tore up the anonymous death threats. On July 15, after hearing from wife of Li Gongpu, Zhang Manjun, the details of her husband's assassination, you were filled with righteous anger and slammed the table, delivering your now-famous speech fiercely denouncing the reactionaries:

> …
>
> Now Mr. Li has been assassinated by the reactionaries for fighting for democracy and peace …
>
> You may kill one Li Gongpu, but millions more will rise in his place!
>
> The power of the people will triumph, and truth will always prevail! …
>
> To fight for democracy and peace, we must be ready to pay the price. We are not afraid of sacrifice. Each of us, like Mr. Li, must walk out the door prepared never to return.

Wen Yiduo, you fulfilled that promise. Your blood was shed in China's fight for the light. You stepped out from behind your scholar's desk and stood at the forefront of the democratic movement. You called out with passion, helping countless people see through the darkness of Kuomintang rule and unite around the Communist Party, launching a mighty movement for democracy that powerfully supported the People's liberation war and helped bury the Chiang regime.

■ Neighbors

The martyrdom of Wen sent me into deep reflection and memories of the past.

I had known him as early as our days at Tsinghua University, though back then, I was merely one among the many young people who admired his brilliance from afar.

It was no coincidence that so many young people looked up to him. In "Dead Water," he likened the darkness of China to a "ditch of hopeless dead water," and in "Heartbeat," he cried out, "Who cares for peace in a walled square inch? My world has broader frontiers …" His words deeply resonated with a generation.

So when I had the chance to live under the same roof as him, even though life was hard and austere, I felt fortunate. In the spring of 1938, our school relocated to Yunnan. To escape Japanese bombings, he moved with his family to Chenjiaying in northern Kunming. My own family, with nowhere to go, also ended up there. He kindly offered us one of his rooms. His family of eight lived in the other two connected rooms. A simple curtain between our households began a shared life that neither of our families would ever forget.

It was in this humble arrangement that I truly came to know him. There, I witnessed how this elder intellectual, shaped by the trials of semi-feudal, semi-colonial China, steadily grew into a brave and unyielding democratic warrior.

During our time living under the same roof, I was bent over my math work, and he was immersed in the study of "Panhu."[1] His modest lifestyle and meticulous academic spirit left a lasting impression on me.

Whether in the chill of spring or the sweltering summer heat, he was always fully absorbed in his research. At night, he'd work by a small oil lamp until deep into the night, immersed in the aroma of ancient books. At that time, he was still on the path he described as "inward development." His study of "Panhu" culminated in a lengthy piece called "A Study on Fuxi," and he often expressed joy in his discoveries.

At the time, few could understand the value of his digging into ancient texts. But later, I saw how Guo Moruo held his work in high regard, praising his reconstruction of myth and legend and saying that he was "revolutionizing Chinese literature from within." Guo noted that Wen's scholarship was driven by a desire to transform thought itself. Through mythology, Wen explored the origins of our people and culture, ultimately achieving internationally recognized accomplishments. That said, it's undeniable that during this period, Wen's interest in "Panhu" clearly outweighed his interest in politics. This shared period of hardship and daily interaction deeply influenced me. Wen's rigorous academic style became a lifelong model for my own studies.

But as the Chinese nation stood on the brink of life or death, and the democratic movement in Kunming surged, Wen, as a just and patriotic intellectual, finally stepped out of the study. I remember clearly—during the May Fourth commemoration in 1944, faced with the despicable performances of reactionary students and faculty, Wen was filled with indignation. He stood up courageously to support the progressive youth. From that moment on, he began to transform from a poet and scholar into a warrior crying out for democracy and peace.

At the time, we were still living in Chenjiaying, but Wen had moved to Kunhua Junior High School in the western part of Kunming. I once spoke with him about

1 Panhu is a general term referring to Wen Yiduo's work at the time on reconstructing ancient myths and legends. It originates from a mythological tale in ancient Chinese lore concerning the origins of humankind.

the changes I saw in him. He became impassioned and said: "Some say I've become extreme, even that I joined the democratic movement out of poverty. But haven't we witnessed with our own eyes how the country has fallen into such a wretched state, how the people live in such suffering? Are we not entitled to a shred of justice? If we don't speak up for justice, we're shameless and selfish!" He went on earnestly: "If not for the displacement we've experienced these years, how could we understand so deeply the suffering of the people? How else could we have seen so clearly the rottenness of the reactionaries?"

This is how Wen broke free from the cloistered world of academia and came to stand with the people, sharing in their suffering. He cast aside all hesitation and plunged into the torrent of the democratic movement.

For this chapter of life, I once wrote a few lines of humble verse:

A curtain split the rooms that barely fit us,
Yet shared hardship bound our households.
East of the cloth, ancient myths were studied; West of the cloth, math problems solved,
Though our subjects differed, our hearts stood united in resistance.

■ The Huize Struggle

In 1944, victories from the front lines of the War of Resistance Against Japan were coming in one after another. On the eve of July 7, the Party decided to hold a political evening forum at Huize Hall (also known as Zhigong Hall) of Yunnan University to commemorate the seventh anniversary of the war. This was the first large-scale public political gathering in Kunming since the Wannan Incident. It aimed to push the surging democratic movement to a new climax, and as such, it aroused the vigilance of the reactionaries, who did everything they could to sabotage the event.

The director of student affairs at Yunnan University declared that only academic topics were allowed at the event, strictly forbidding any political discussions. Military police were dispatched to "maintain order." Despite the gathering of three to four thousand people, no sound equipment was provided. The reactionaries even planted troublemakers among the audience to stir up disruption.

University President Xiong Qinglai, under orders, gave a long-winded speech at the event. He deliberately avoided the purpose of the evening—to commemorate the War of Resistance—and instead spoke extensively in abstract numbers. He especially emphasized the idea that "change" is wrong and could lead to chaos. Clearly, this rhetoric, dictated by others, was detrimental to the democratic movement.

Wen, angered by Xiong's remarks, requested permission from the chairperson to speak. The audience immediately erupted in enthusiastic applause. He then delivered a passionate and stirring speech that electrified the entire Huize Hall: "... Some don't like this gathering and don't approve of discussing politics. They say it's not a matter for us educators." "The speakers tonight are all old colleagues and friends, but since our opinions differ, I feel the need to speak up ..." "The country is in such dire straits—if we don't speak out now, when will we? If we don't take responsibility, who will?"

Wen was deeply moved as he spoke. He powerfully refuted the argument that "students should just study" with his own belief that "students should care about national affairs," which resonated deeply with the crowd. This debate defended the democratic movement, and his speech had a wide and lasting impact. He made a significant contribution to the development of the democratic movement.

Afterward, Party members in the underground asked me to talk with President Xiong to persuade him and prevent "Yunnan University"—this stronghold of democracy—from falling into the hands of the reactionaries. President Xiong told me, "It was the director of student affairs who sent me. I was deceived by the secret police. I shouldn't have gone. If you see Wen, please help me explain." Later, I relayed Xiong's words to Wen, who responded with understanding: "It had to be done at the time. Admittedly, my speech was a bit too sharp."

The Huize Struggle was just one episode in the larger democratic movement in Kunming, but its influence at the time was profound. It not only dealt a blow to the arrogance of the reactionaries, but it also awakened many intellectuals who had been wavering, thereby strengthening the democratic front. The Huize Struggle was also a triumph of the Party's united front policy—it embodied the brilliant strategy of uniting all possible forces in common struggle, including those with confused or erroneous views.

Under the leadership of the Communist Party, the China Democratic League did extensive work in uniting the broad masses of intellectuals. However, during the "Cultural Revolution," the villainous "Gang of Four" demonized democratic parties, branding them as monsters and demons, while proclaiming their own exclusive revolutionary legitimacy. This caused catastrophic damage to the Party's united front. These contrasting lessons—both positive and negative—help us to better understand today the vital importance of cherishing the Party's united front, which is a powerful tool for unity, resistance, and ultimate victory.

■ Carving Seals in Commemoration, Encouraging the Next Generation

Wen had a clear sense of love and hate—he glared fiercely at the reactionaries, but for his friends and comrades, he held deep affection.

To this day, I still treasure a seal that he gave me. It has a few lines of small characters carved on it: "A stubborn piece of stone, carved by Yiduo. Presented to the professor, as payment and pledge. Not shabby, not extravagant. Crude compared to ivory seals, yet elegant as a wooden stamp. Before the war, it wouldn't be worth even twenty cents."

This simple stone seal, combining elegant calligraphy with exquisite carving, provides a deeply satisfying artistic experience.

While in Yunnan, Wen held positions both as a university professor and a high school teacher, yet still struggled to feed his family of eight. Starting in the summer of 1944, in order to ensure the bare minimum for his household, he took up a new profession—becoming a craftsman, and began carving seals.

His father was a licentiate and with a strong foundation in traditional learning, combined with his early specialization in the arts, he had a unique talent in engraving and design, achieving a high level of artistry. Yet he never imagined that one day, for the sake of survival, he would have to publicly hang a sign offering seal-carving services.

A sign bearing "Wen Yiduo Carves Seals" appeared in the streets of Kunming. Well-meaning colleagues even wrote an introductory blurb for him: "... A pioneer in literature, a renowned scholar in classical studies. He discerns script down to the finest strokes—how many can truly appreciate it? He speaks of the origins of elegance and art—respected throughout the land."

Customers came in an endless stream. Some bureaucrats with a taste for elegance even sent ivory to Wen for carving. But he firmly returned them all, never tempted by the promise of profit.

Wen carved seals for a living, but he meticulously crafted one as a gift to me. It stands as a perfect artistic keepsake, and more importantly, a treasured token of his sincere friendship. Throughout the many years of wandering and upheaval, I have always kept it with me. Every time I take it out, I think of him. The revolutionary camaraderie embodied in that seal inspires me to keep pushing forward.

In September 1945, Wen was elected as an executive member of the Central Committee of the China Democratic League. As he drew closer to the Party, he also made great efforts to unite his friends and encourage them to move forward together. He passionately supported every progressive action taken by his friends.

In March 1946, the Soviet Academy of Sciences published my mathematical work, and soon after, they extended an invitation for me to visit the Soviet Union. When Wen heard the news, he enthusiastically encouraged me not to fear the difficulties of the journey, to break through the sabotage and obstruction of the reactionaries, and to resolutely go abroad for research and study. He said earnestly: "We must learn from the Soviet Union and follow the Soviet path. Your opportunity to study there will greatly benefit China's future in science. You absolutely cannot miss this chance!" With his encouragement, I finally overcame numerous obstacles and visited the Soviet Union.

After returning to China, the underground Party organization at National Southwest Associated University asked me to give a report on my visit to the Soviet Union. The talk was originally scheduled in a Youth Association classroom, but due to overwhelming attendance, it was moved to the large athletic field. To guard against possible assassination by secret agents, the underground Party arranged for me to speak from a balcony. The report was a success. Wen praised me, saying: "You gave a very thorough introduction to the situation in the Soviet Union. It was excellent—and very beneficial to the current democratic movement." I never expected those heartfelt words would become his final words to me—his last encouragement.

It has been over three decades since Wen was murdered. In those decades, China has undergone earth-shaking changes. The ideal of a new China that Wen devoted his life to has become a living reality under the leadership of the Party and Chairman Mao.

We must learn from Wen's spirit—his courage to step forward and sacrifice himself for the interests of the country and the people at a critical moment when China faced two opposing futures and destinies. He will forever remain a shining example for Chinese intellectuals.

Today, our people, united around the Party Central Committee, are marching toward socialist modernization. Achieving the Four Modernizations and building China into a prosperous, strong socialist country is the very ideal that Wen and countless martyrs fought and died for. We will bring this ideal to life with our own hands. Our task is both glorious and arduous. As a younger peer and friend of Wen, I have always felt ashamed. In the darkest times, I did not step forward as he did, to trade my life for the dawn. But now I take comfort in the fact that I can dedicate the rest of my life to completing the unfinished work of Wen and so many predecessors. As Premier Zhou once said: "Lu Xun and Wen Yiduo were the most loyal and diligent oxen. We must learn from their example." On the road to socialist modernization, we must carry forward the ox-like spirit of Lu Xun and Wen Yiduo—working tirelessly for our country and for our people.

In closing, let me offer the following two short poems in honor of Wen's 80th birthday:

You struck the table in righteous rage—
I fled in fear, a coward's stage.
Awakened late, I mend my shame,
White-haired, I carry the martyr's flame.
My white hair I offer to modernization's cause—
Not resting until the people's hardship draws pause.
For the Party, I'll fall wherever the fight may swell—
Be it riverbanks or a barren dell.

Some Reflections on Studying and Researching Mathematics*

It is important for a person to have a clear understanding of themselves. I am aware that my grasp of scientific research is not comprehensive. I also know that a crucial aspect of doing science is independent thinking. Everyone should find the path that best suits their own characteristics. Simply hearing about someone else's method for learning or conducting research and assuming that I too can study and research effectively—that would be no different from thinking one can become an immortal just by swallowing a magic elixir, without any hard cultivation or practice.

Today, I'd like to share with you my experience and lessons learned over the past fifty years—what I've seen, heard, and personally understood. My purpose is to offer my experiences as reference for you. Ultimately, each specific person must find the most effective method suited to their own situation through specific analysis of specific problems.

■ Speed and Efficiency

Speed is a guarantee for achieving socialist modernization. Let me give a simple example: I'm old and slow now—if I walk ahead and you follow me, it's easy for you to catch up. But if I try to catch up with you, unless you lie down and take a nap, there's no way I can reach you. Scientific development in the world today is

* This article is a speech delivered to the graduate students of the University of Science and Technology of China. It was first published in issue 1 of *Mathematical Bulletin* in 1979.

moving very fast. If we don't exceed the speed and efficiency of the United States, we won't be able to catch up. If we don't exceed the speed and efficiency of Japan, we won't be able to catch up with Japan either. If our speed merely matches that of countries like the US and Japan, then we'll only be closing the time gap—they'll always remain ahead, and surpassing them will just be empty talk. Therefore, we must first surpass them in speed and efficiency.

But is it even possible for us to surpass them in speed and efficiency? This might seem like a big issue—but in reality, it's not. I've lived in the US, in the UK, and also in the Soviet Union. I've seen firsthand that their speed is not some unreachable myth. We can definitely catch up—and even surpass them! Many Chinese scientists working in America became renowned in modern science precisely because their speed and efficiency surpassed the average American scientist. So, facts prove that if we put in the effort, catching up and surpassing them is entirely possible.

Take myself, for instance. I went to Britain in 1936 and stayed there for two years. After returning to China, I lived in the countryside near Kunming for two more years. By 1940, I had completed my work on the theory of sieve methods in prime numbers. After returning to China again in 1950, by 1958, our research in number theory, algebra, and the theory of functions of several complex variables had already reached an advanced level on the global stage. Experience tells us that within four or five years, a branch of pure mathematics can begin to stand out globally.

You are living in a much better era. The Central Committee has crushed the "Gang of Four," bringing a springtime for science. In these favorable conditions, I can confidently say: As long as we work hard and study diligently—not wasting a single second—we can absolutely reach the world's leading level. This is especially true in mathematics. We have pioneers such as Xiong Qinglai, Chen Jiangong, and Su Buqing as shining examples from the past. And today, many promising young talents are already emerging, with even more certain to follow.

■ Digestion

Striving for speed does not mean skipping levels or jumping around blindly. It doesn't mean moving on to another book before fully digesting the one you're reading, or switching to a new field before having climbed to the peak of your current one. Our studies must begin with steady, grounded reading.

In ancient times, people often praised someone as being "well-read and possessing a powerful memory," or as "having knowledge to fill five cartloads of books." But many of these sayings from the past are no longer reliable today. The five cartloads of books in ancient times were large-print volumes—nowadays, just a fingernail-sized integrated circuit could store five or even ten such books. So "knowledge to fill five cartloads" today amounts to just a few chips. Likewise, people now may say, "So-and-so has read countless books," or "So-and-so remembers the world's literature in great detail." Of course, this isn't useless—it just represents the very first step. If you don't digest what you've read, then you're no better than a library or a computer memory system.

What makes humans valuable is our ability to create—to absorb the essence of past literature and, through digestion, produce what did not exist before. Otherwise, if we just echo others mindlessly, the world would never progress. Lazy thinking is the enemy of science—and naturally, of societal development too.

What does it mean to "digest" knowledge? The best way to test whether knowledge has been digested is through use. Whether you can use it or not is not determined by empty talk, but by real practice. If you're helpless when faced with a problem, or clueless when confronted with another, then even if you've written a few imitative articles or compiled some translated works, what relevance does that have to society's progress? Of course, I'm not against beginners writing some imitative articles—but one cannot stop there. If a person grows old and grey and still only copies others, that's not true learning.

Only after digestion can we apply knowledge flexibly. When socialist construction needs us, we'll be able to serve it, solve problems, and contribute our strength. Real-world problems don't come with labels telling you, "This one requires number theory," or "That one needs functional analysis." The challenges of socialist construction are varied and endless. Trying to apply one single method to all real-world problems—that's metaphysical thinking. Only those who think independently and digest thoroughly can choose and apply the right methods, tailored to different situations and problems.

Now, as I just said, digestion is ultimately tested in practice. But students don't always have many real-world opportunities. So, is there a way to check, while still in school, whether we've truly digested what we've learned?

I've said before that learning follows a process of "from thin to thick, then from thick to thin." When you first study a book, you annotate it heavily and read many reference texts—what was a thin book becomes thick. You might think, "Now I understand it!" But that would be self-deception. That's not yet true understanding. True understanding involves a process of turning "thick" back into "thin." After you analyze the whole book, discard the trivial bits, grasp the key points, and even see the entire thread clearly at a glance—only then can you say you're beginning to understand. Try to think: Before this theorem existed, how did someone come up with it? That's also a way to test whether you've digested it. Admittedly, this method isn't as solid as real-world practice. But the overall principle remains: The ultimate test of digestion and comprehension is practice—whether you can flexibly apply what you've learned to solve real problems.

Some may say, "But learning this way is too slow." My own experience says otherwise—it's actually faster. Because when we've digested what we previously studied, our memory system automatically filters out familiar content when reading a new book, and focuses only on the new material we haven't seen before. So, this grounded approach is not slow, but fast.

Otherwise, if you learn in a rushed and shallow way, you'll just forget things again. And when you study, forget, then study and forget again, time is wasted—which is a small loss—but worse, you make yourself "useless to the country and to your family," which is a huge loss.

Even more dangerous is being overly ambitious and aiming too high. For example, if someone hasn't fully grasped junior high school math but is already tackling junior or senior university-level courses, they'll struggle, but still feel it's beneath them to go back and review the basics. Continuing in this way only leads to increasing confusion, like sinking into a swamp with no way out.

Sometimes, when reading a book of the same level, if we've truly understood and digested previous texts, we can quickly pick out the new points that weren't covered before. If nothing new is found, we can shelve the book quickly. If we find one or two new things, we just need to focus on understanding those—and that's it. That's how reading becomes fast, not slow.

Once you've learned how to read effectively, you can move on to reading research papers. In essence, reading literature and reading books are not that different—they both require digestion. However, books tend to contain more refined, mature knowledge, so when you sift through the content, there's more essence than dross. Academic papers, on the other hand, are often fresh off the press, and typically have more dross than essence. Of course, some papers immediately become classics—but those are rare exceptions. The majority of papers are forgotten not long after they're published. Once you've built a solid foundation for understanding research literature, you're ready to do research.

At this point, I want to emphasize independent thinking. It is the foundation of scientific research. Throughout history, not a single major invention was created without it. That said, independent thinking doesn't mean ignoring the achievements of your predecessors. For example, some people dive into studying the Goldbach Conjecture without any knowledge of prior work—that kind of approach has little chance of success.

Independent thinking also doesn't mean refusing to study books, read papers, or listen to teachers. Books, literature, and teachers are all necessary—but if you rely on them too rigidly, you'll lose your creative spark, and students may end up becoming mere extensions of their teachers. This makes your intellectual world grow smaller and smaller, and in mathematics, this leads to what I call a "convergent phenomenon." Only through independent thinking can we break out of that box, invent new methods, explore new fields, and push science forward.

Independent thinking doesn't mean sitting alone and brooding in silence without exchanging ideas. It still involves building on others' results and drawing on collective wisdom. It's also a way to compensate for shortcomings in our mentors. If your mentor lacks experience or is too busy to guide you properly, then independent thinking becomes all the more essential. In fact, back when we were in Kunming during wartime blockades, we had no access to foreign journals. Still, we managed to think independently and come up with new ideas—ideas which, as it turned out, didn't overlap with what was being done abroad. And even if they had, so what?

For example, when I was young, I published a few papers from home, and received many rejections. The reasons? Editors would say, "Your theorem is already in such-and-such a book," or "This has already been proven in that article," and so on. Should I have given up? Should I have thought I wasn't good enough, or that I was just wasting time repeating what others had done? At the time, I didn't think that way. In fact, I was glad when I received those rejections—it meant that what a top scientist had discovered, I was able to discover on my own in a humble setting. That encouraged me to double down and keep going.

Let me be clear—I'm not saying we shouldn't read the literature. What I am saying is: Even if we accidentally repeat someone else's work, we shouldn't get discouraged. Instead, compare your result with the existing one. See how they differ. Are theirs better? Then learn from their strengths. If yours has strengths too, then you've gained confidence. Either way, it's progress. If we think independently, then a lack of a mentor or incomplete literature won't hold us back. On the contrary, even when we do have mentors, we still need to evaluate what they say. Are their ideas correct? Is the literature comprehensive?

In short, science is built by those who can think independently—not by people who are like memory chips, storing data without creativity. Such people cannot innovate. Chairman Mao once said that in research we must discard the dross and extract the essence, separate the false from the true, move from the superficial to the deep, and move from the particular to the general. To truly achieve those four things, independent thinking is essential. Without it, all you'll get is surface-level understanding, crude conclusions, and an inability to distinguish truth from falsehood.

■ Several Levels in Research

There are several levels in research:

1. Imitating by following a model step by step. Work based on imitation is essentially like doing exercises. Of course, doing exercises is necessary, but what's the point if one spends a lifetime doing exercises without any innovation?
2. Using existing methods to solve new problems. This is a step forward compared to simple imitation. But even here, we should distinguish further: Directly applying a known method is still similar to doing an exercise. However, if one applies an existing method with some modifications, then this begins to resemble true scientific research.
3. Creating new methods to solve problems. This is an even greater advancement. The creation of new methods marks a major turning point and demonstrates significant improvement in one's abilities.

4. Opening up a new direction. This is a higher level still. If one opens up a new direction, it could lead others to work in that field for decades or even centuries. This kind of contribution has a deep and lasting impact on scientific development.

These are rough categorizations, and mathematics has many of its own particularities. For instance, improving or solving a famous problem is often highly praised in mathematics. At the beginning of the 20th century, Hilbert proposed 23 famous problems. Some of these have had an enormous influence on the nature of mathematics itself.

Take Fermat's Last Theorem, for example—most people know about it. If one were to solve it using only elementary number theory, the result might earn the prize of "100,000 Marks," but it wouldn't contribute much to the development of mathematics. On the other hand, Kummer, though he didn't solve Fermat's Last Theorem, created the concept of ideal numbers during his study of it—thus opening up an entirely new direction. Today, this concept is used in algebra, geometry, analysis, and many other fields. So, his contribution far surpassed that of solving just the single problem of Fermat's Last Theorem. I believe this kind of contribution is greater than solving isolated difficult problems.

I would like to give everyone a piece of advice: "Aim for the highest, and you'll at least reach the middle; aim only for the middle, and you may fall below." There's another principle in research worth noting: One must be able to break in and also break out. Breaking in requires theory—you must penetrate deeply into the core of your chosen topic. This is well understood. However, breaking out is not necessarily easier. Many mathematicians have managed to dig deep into a topic but then found themselves stuck in a dead end, unable to get out. This often leads to being trapped in a narrow issue for years, wasting time. While this may bring personal satisfaction, it doesn't contribute meaningfully to science or to the needs of socialist construction.

■ About Extension

When moving from one branch to another, should we completely abandon the original branch and jump directly to another? If so, we risk forgetting everything we previously learned. The idea of extension is to gradually expand our vision from the field we've already mastered—spreading outward into related areas. In this process, we transition into new fields while still making use of our prior knowledge. This expands our range and makes us more adaptable. I support those who spend many years digging deeply into one problem and producing solid results. But I also support the idea that, after achieving results, one can slowly extend into other fields using this approach of extension.

When assessing a scholar or individual, one must consider both breadth and depth. Depth alone may make a solid expert, but their impact on the broader development of science may be minimal. Breadth alone—knowing a little about many things—may impress laypeople and show off versatility, but it cannot produce meaningful results for the people.

Between various branches of mathematics, and between mathematics and other disciplines, there are no true unbridgeable gaps. In the past, we often saw strict compartmentalization and everyone working in their own tiny area. What was the result? None of those areas developed well. Therefore, while specializing in one field, it's highly beneficial to read books and literature from related areas.

■ About Rigor

Rigor is not only essential for scientific research—even for military training, it must be rigorous and demanding. It's always nice to hear pleasant words, and people are naturally happy when others speak kindly to them. Back in the Three Kingdoms period, there were two men: Kong Rong and Mi Heng. Mi Heng flattered Kong Rong, saying he was Confucius reincarnated; Kong Rong praised Mi Heng as Yan Hui reborn. Although they praised each other to the skies, in reality they were both good-for-nothings. In the end, both were either directly or indirectly killed by Cao Cao.

Of course, hearing flattery feels good, and some people who offer compliments argue that they are "encouraging the youth." But this kind of encouragement sometimes has the opposite effect—it inflates young people's egos and hinders their growth. In recent years, I've seen many teachers who held their students to high standards get criticized, while others who were lenient with grades and focused on "maintaining relationships" were more popular among students. This kind of trend only drags back socialist progress. As a result, even when we want a teacher to be strict, they're afraid to be truly demanding. That's why I hope students will take the initiative to ask teachers to be strict with them, and for teachers who are reluctant, we should engage them in conversation to help ease their concerns.

After all, saying a few strict words is just as easy as saying something flattering—only people tend to prefer praise. Many even think compliments are a sign of unity, and as soon as they hear criticism, they assume there's a lack of unity and feel that someone needs ideological counseling. But that's redundant. A strict teacher-student relationship only strengthens unity. Even if there are temporary misunderstandings, over the long run of study and research, students will come to appreciate the value of a strict teacher.

Meanwhile, we must also be strict with ourselves. The style of the "Three Honests and Four Stricts" from Daqing is something we should follow at all times, whether in public or private.

Previously, I talked about the idea of "digestion"—this is a reflection of rigor and self-discipline. Reading a book carelessly might get you by in school, but if you graduate like that and enter the workforce, you'll find yourself unprepared. I also have had a personal lesson about rigor. In 1964, when I was just starting to apply theory to practice, I was working in a remote area described as "challenging but intimidates none." An engineer, concerned about wasting state resources, told me: "The yield rate for detonators is low. Can you lower the standards a little so more of them can pass inspection?" At the time, I thought this was easy—I just needed to slightly lower the standards, and the acceptance rate would rise. But later, I had a deeply painful lesson at Meihua Mountain. I came to understand that lowering the standards by just 1% could mean the lives of four of our beloved soldiers. That was the beginning of our work on the optimization. Once a defective product has been made, even if we catch it during inspection, the loss has already occurred. But if we improve the quality from the beginning, reduce defective products, the damage to the country is naturally reduced.

This isn't to say that quality control isn't important. In fact, I advocated for it as early as 1969. But the core of optimization is prevention—like treating and preventing illness—prevention should take priority. Implementing optimization is about preventing defective products from appearing in the first place. In contrast, "treatment" means reworking the defective products after they've already been made—but this often can't be done, turning into an incurable problem. To be honest, I used to be lenient with students who made small calculation errors in exercises. But ever since that painful experience, I've been deeply educated. A 1% error in an office setting may seem trivial, but once it enters real-world production or construction, it can lead to enormous losses.

So, in short, we should be immensely grateful to those who hold us to high standards—myself included. As for those who flatter us, sure, we can thank them too—but I'll return that flattery and let them keep it for themselves.

Students, learning is like rowing upstream: If you don't move forward, you fall back. The moment we stop holding ourselves to strict standards, problems will arise. Of course, every mathematician has made mistakes. I'll say it like this: Only a mute person has never said the wrong words, only a fool has never thought the wrong thought, and there is no mathematician who has never made a mistake in calculation. Mistakes are inevitable, but they should never be an excuse to lower our standards. Our goal is zero mistakes. When errors do happen, we must learn from them.

Empty bragging may feel good, but we must learn to analyze things clearly. Stay respectful but cautious of those who constantly flatter—there's an old saying: Everything can be exposed, except flattery. Irresponsible flattery is a harmful habit left over from the old society, and we must eradicate it. Of course, when others say good things about us, we don't need to argue back—but we must remain coolheaded and clear-minded, recognizing that these compliments are worthless flattery that do nothing to help our growth.

Seeking truth from facts is the foundation of science. If those engaged in scientific work are not truthful, they cannot succeed in science or may not be suited for it at all. The Party has repeatedly advocated for the spirit of seeking truth from facts. Those who speak or act without truthfulness go against the Party's requirements. Science does not tolerate any falsehoods. We must correctly assess what is truly good—if it does not receive recognition at first, do not be discouraged because practice will prove its value. On the other hand, things that seem good at first but are not truly so will soon fade away, no matter how much praise they receive initially.

We must have perseverance and the ability to persist, but when we realize we are in a dead end, we must also be able to change direction. However, it's not easy to realize when we've reached a dead end—it requires hard work and effort. Even when we retreat from a dead end, we must understand why it is so. After several years, if the obstacles are cleared and the challenges are overcome, we can reenter it for another go. Failure is common, and success is often accidental. All published achievements are successful outcomes that we see, but what we don't know is that these successes are the result of countless failures and lessons learned.

The advantage of learning from a good teacher is that they can guide us to avoid failures and help us absorb successful experiences more quickly. This foundation allows us to create even better things. Moreover, we can also learn from their failures, understand how they overcame obstacles, and see how they turned failures into success. Never adopt the mindset of hoping for success without putting in hard work. True talent involves a process of countless failures before solving the problem elegantly. So, students, do not fear failure. Never think that because you've written a hundred pages and still failed, or because you've worked on a problem for two years without a result, that you should lose confidence. Instead, we should summarize our experiences, identify the reasons for our failures, and avoid repeating the same mistakes. In short, failure is the mother of success.

To appear to understand when you don't, or to pretend to understand when you don't, is even worse than not understanding at all. Such people have no future in scientific research and are blind guides in scientific management. If you truly know yourself and admit what you don't understand, you will be more willing to listen to the opinions of others, analyze those opinions, respect the expertise of others, and then make decisions collectively. Especially for young people who have not yet undergone the test of hardship (the hardest test being the test of war, where poor judgment leads to defeat and even destruction), they don't yet know the struggles of others. Without the ability to organize mass scientific research, they will inevitably fall into the trap of mismanagement. Even though such individuals may have great ambitions to advance science, in reality, they may end up sabotaging progress.

Therefore, I ask you young people to develop two key abilities: To cultivate deep and thorough expertise in one or two scientific fields. To be capable of organizing scientific experiments as mass movements—mobilizing the public and spreading scientific knowledge among the people. Otherwise, it will only hold back the advancement of the Four Modernizations. As for the individual, they will achieve nothing, and by then their hair will already have turned gray.

When these two paths seem incompatible, choose one, but remember, there is no such thing as a bountiful harvest without working the fields, nor producing goods without working beside the machines. Mental labor is the same. A scientist or science organizer who is idle, well-fed, and inattentive to their work will not succeed.

No scientist can rely solely on innate talent. Only through hard work can diligence overcome weakness and unlock true potential. Those with less natural talent, through effort and perseverance, can catch up with and surpass those who have talent but lack hard work. As the ancients said, "If others can master something with one effort, I will do it ten times. If they succeed with ten efforts, I will strive a hundred times." This timeless wisdom is highly valuable.

■ Be Willing to Expose Yourself

Is it good to pretend to understand something when you don't? No! Because if you pretend to understand, you will never truly understand. You must dare to expose your weaknesses and areas where you lack knowledge. Don't be afraid of feeling embarrassed. The worst that can happen is that a teacher might reproach you, saying, "You don't even understand this?" But after being reproached, won't you understand? However, if you refuse to be criticized and pretend to understand, you will never truly do.

Science is based on truth and facts. The more knowledgeable a person is, the more they are willing to expose their limitations. They can say, "I don't understand this part," and through discussion, they will come to understand it. In broader terms, the flourishing of diverse schools of thought is exactly this: Each school dares to expose its own ideas, critique others' ideas, and accept the strengths of others. This is how science thrives and prospers. During the era of the "Gang of Four," everyone was discouraged from openly stating their positions. Not only was it bad to remain silent, but even when something was clearly wrong, people were afraid to expose it.

This led to rigidity, and rigidity is the archenemy of science. Without openness and exchange, scientific progress comes to a standstill.

Don't fear being humble; fear not knowing your own limitations. When you expose your weaknesses, teachers can locate where you need help and can teach you accordingly. At the same time, don't pretend to not understand something when you do. If the teacher knows that you understand a lot, they can help you progress faster. In short, when you understand, admit you do; when you don't, admit you don't; when you can do something, admit you can; when you can't, admit you can't. This is the scientific attitude.

Good performance may seem like a bad thing, but it deserves analysis. If you don't truly understand something, or if you only have partial knowledge but show off your supposed expertise, that's truly bad performance. On the other hand, if you share your understanding with others so they can learn from your strengths in less time, that's not just "showing off"—that's good performance. Science depends on mutual contact and exchange of ideas, and this is how we can thrive and prosper.

What I've discussed above may be one-sided, and what's more important is the issue of serving the people. Your political theory studies are probably better than mine, and since we're out of time, I won't go into it further. I'll conclude with one sentence: Serve not for yourself, but for the people.

Of course, my speech has this theme, but I haven't been able to fully elaborate on it. However, one must have self-awareness. My understanding in this area is weaker than my understanding of mathematics, and political leaders are likely to know more about this than I do. So, I won't go beyond my area of expertise and speak too much on it here.

Some Personal Experiences in Popularizing Mathematical Methods in the People's Republic of China*

Introduction

At the Fourth International Congress on Mathematical Education, I had the honor of being one of the four keynote speakers. Personally, I feel this is a great honor—not just for myself, but also for the Chinese people. On the other hand, it is important to know one's own limitations. I am largely self-taught in mathematics, and I have relatively little direct experience in mathematics education. Over the past twenty years, I have focused on bringing mathematical methods to workers and technical personnel in order to serve production. My work has been a combination of theoretical research, teaching, and hands-on exploration in practical settings.

This is my first opportunity to learn from the world's leading mathematics educators. In the field of mathematics education, I am still very much a beginner and a self-learner. I have not yet studied many of the renowned works in this area. If there are shortcomings or mistakes in my speech, I sincerely welcome your corrections and guidance.

---

* This is the revised version of Comrade Hua Luogeng's report presented at the 4th International Congress on Mathematical Education. For the original English text, see: L. K. Hua and H. Tong, "Some Personal Experiences in Popularizing Mathematical Methods in the People's Republic of China," *Int. J. Math. Educ. Sci. Technol.* 13, no. 4 (1982): 371–386. This congress was held in August 1980 in the city of Berkeley.

■ Three Principles

Starting from the mid-60s, apart from continuing my theoretical research, I was actively involved in popularizing mathematical methods to the general public in China. Altogether, I and my assistants have visited twenty-two provinces, hundreds of cities, and thousands of factories, meeting millions of people. Such was the magnitude of my personal involvement. To sum up experiences accumulated in this still ongoing course, the following principles will form the basis of the discussion:

1. To whom? / For what purpose?
2. Which technique?
3. How to popularize?

Some elaboration of these three principles is now in order:

To whom? Specialists or experts and factory workers do not often share a common language. My experiences have shown me that in order to achieve a common language, these two groups must look for their common need. Needless to say, it is unrealistic to expect the factory workers to be interested in infinite dimensional spaces, although these objects are important for the mathematicians! The worker hopes to quickly find the balance point of the grinding wheel or the cylinder. Therefore, when carrying out educational outreach, the first step is to identify a common goal between the speaker and the audience. With a shared goal, it becomes possible to develop a shared language. Only then can we move on to the next question.

Which technique? Regarding this question, I will elaborate more later. For now, I would like to introduce the "three principles of topic selection":

1. Mass appeal. The methods we introduce must be understandable, learnable, and applicable by the relevant groups, and they must produce tangible results.
2. Practicality. Every method must undergo practical testing before being widely promoted. Practice helps us define the scope of applicability, so that popularization can take place within this scope. Through experience, we have found that methods successful abroad may not necessarily yield the same results when directly transplanted to China.
3. Theoretical foundation. A strong theoretical foundation is essential. Theory

enables us to explain complex ideas in simple terms, distinguish between effective and ineffective methods, and develop new techniques.

How to popularize? My experiences have shown me that it is essential to start work on a small scale, for example, in a workshop. If my suggestions turned out to be effective they would naturally attract wider attention and might soon spread to other workshops, and then to the whole factory or even to a whole city or a whole province. In this way, my assistants and I sometimes had more than a hundred thousand people in the audience! The lecture format involved one main venue along with several breakout rooms. Since closed-circuit television was not yet widely available, each breakout room had one of my assistants present to handle demonstrations and illustrations. After the lecture, our responsibility was not only to answer questions, but more importantly, to go to the work sites, engage in hands-on practice with everyone, and ensure that the methods we taught produced real, effective results in actual production.

▪ Searching Through Books

As a scholar, one often turns to literature or books in search of material. If one pays attention to analysis and comparison, this approach can indeed be a good method—it allows for the gaining of valuable experiences and lessons. There are many examples of this; I will mention just one.

How to find the surface area of a mountain? We found two methods from the books. A method often used by geologists is the so-called Бауман method and that by geographers the so-called Волков method. We may briefly describe these two methods in the following way:

Starting with a topographic map that has contour lines with an elevation interval of Δh. l_0 is the contour line at elevation 0, l_1 is the contour line at elevation Δh, …, l_n is the highest contour line, at elevation h.

Let W_i be the area between l_i and l_{i+1} on the horizontal plane.

Бауман Method: This method has two steps:

1.

$$C_i = \frac{1}{2}(|l_i| + |l_{i+1}|)$$

where $|l_i|$ denotes the length of the contour line l_i.

2.

$$B_n = \sum_{i=0}^{n-1} \sqrt{W_i^2 + C_i^2}$$

Geologists take B_n as the estimated surface area of the mountainous region.

Волков Method: This method also has two steps:

1.

$$l = \sum_{i=1}^{n} |l_i|,\ W = \sum_{i=0}^{n-1} W_i$$

2.

$$V_n = \sqrt{W^2 + (\Delta h \cdot l)^2}$$

Geographers take V_n as the estimated surface area of the mountainous region.

These two methods, coming from different scientific disciplines, naturally raise two important questions when presented side by side: Do they converge to the real surface area? If so, which method is better?

Disappointingly, neither method converges to the real surface area A. To be precise, define

$$B = \lim_{n\to\infty} B_n,\ V = \lim_{n\to\infty} V_n$$

Then we have the inequality: $V \leq B \leq A$.

The proof is not difficult, but it is somewhat interesting. We can represent the surface by the equation: $\rho = \rho(z, \theta),\ 0 \leq \theta \leq 2\pi$. This is the contour line at height z, using the peak as the origin. From known formulas, the true surface area is

$$A = \int_0^h \int_0^{2\pi} \sqrt{\rho^2 + \left(\frac{\partial \rho}{\partial \theta}\right)^2 + \left(\rho \frac{\partial \rho}{\partial z}\right)^2}\, \mathrm{d}\theta \mathrm{d}z$$

Introducing a complex-valued function

$$f(z,\theta)=-\rho\frac{\partial f}{\partial z}+i\sqrt{\rho^2+\left(\frac{\partial \rho}{\partial \theta}\right)^2}$$

then we have

$$V=\left|\int_0^h\int_0^{2\pi} f(z,\theta)\mathrm{d}\theta\mathrm{d}z\right|\le B=\int_0^h\left|\int_0^{2\pi} f(z,\theta)\mathrm{d}\theta\right|\mathrm{d}z$$
$$\le A=\int_0^h\int_0^{2\pi}|f(z,\theta)|\mathrm{d}\theta\mathrm{d}z$$

We also discovered that unfortunately the equality is only achieved in very special cases.

This example illustrates two important points: First, it shows the possibility of discovering valuable problems in other fields of science. Second, it demonstrates the role of mathematical theory. Without it, we cannot evaluate the quality of different methods. But with theoretical analysis, we may create better methods.

Once we have found a better method, does it automatically become material that should be popularized? No! This kind of method is best kept among geologists and geographers. That is, we can suggest they adopt this new method when writing their books, or we can use it as a supplementary example when teaching calculus.

Although this is not something suitable for mass dissemination, I still believe such work is necessary. When we have accumulated enough of this type of material, it will make our textbooks more substantial. Exercises will be more practical, and not just abstract conceptual puzzles or artificially constructed problems.

■ Searching in the Workshop

Many meaningful problems can also come from workshops or individual workers. Here's one example I'd like to share—it's called the pulley problem.

It was in 1973 when I and my assistants visited the city of Luoyang in central China on one of our popularization missions. A worker in a tractor factory there presented us with the following problem.

Mathematically, the problem is to find integers a, b, c, and d from {20, …, 100} such that for a given real number, ξ,

$$\left|\xi - \frac{a \times b}{c \times d}\right|$$

is a minimum.

This worker pointed out to us that the values found in mechanical handbooks were not accurate. He took the example of $\xi = \pi$. The handbook gave the approximation:

$$\frac{377}{120} = \frac{52 \times 29}{20 \times 24}$$

but he found a better one himself:

$$\frac{2108}{671} = \frac{68 \times 62}{22 \times 61}$$

which indeed proved to be more accurate. He asked: Is there an even better approximation?

This is a Diophantine approximation problem. At first glance, it seems simple and potentially solvable using continued fractions. Perhaps among the convergent of π such as

$$\frac{3}{1}, \frac{22}{7}, \frac{333}{106}, \frac{355}{113}, \frac{103993}{33102}$$

we could find a better approximation than the one provided by the worker. But no—the fractions before $\frac{355}{113}$ are too rough and no better than his; the ones after have numerators and denominators exceeding 100^2, which does not meet the requirement. Moreover, 113 is a prime number and cannot be factored into $c \times d$. The problem turned out to be quite tricky.

There was only one day left! Before I left Luoyang, I left a small note to my assistant at the train station:

$$\boxed{\frac{377}{120} = \frac{22 + 355}{7 + 113}}$$

Upon reading the note, he understood my suggestion of attacking the problem with the method of Farey means.

Using this method, he found two better approximations:

$$\frac{19 \times 355 + 3 \times 333}{19 \times 113 + 3 \times 106} = \frac{7744}{2465} = \frac{88 \times 88}{85 \times 29}$$

and

$$\frac{11 \times 355 + 22}{11 \times 113 + 7} = \frac{3927}{1250} = \frac{51 \times 77}{50 \times 25}$$

The last fraction turned out to be the best.

While this example uses π, the method derived here can be applied to approximate any real number. Based on this method, we discovered that many entries in engineering handbooks—values for a, b, c, and d—are neither optimal nor complete. I mention this to suggest that we could use such experience to assist the authors and institutions compiling these manuals to improve their quality.

Now, having found this method, does it qualify as a topic for popularization? While more people might need this method than those calculating mountain surface areas, still, only a very small portion of workers actually perform pulley calculations. Moreover, if engineering handbooks are improved, the need for such individual computations is further reduced. Thus, topic selection remains a matter that requires thoughtful discussion and exploration.

■ Optimization

The trial-and-error method is one we frequently use. But what kind of trial-and-error is the most efficient? In 1952, J. Kiefer solved this problem. Because it's related to the golden ratio in elementary geometry, it's called the golden section method, which has a wide range of applications, so how can we teach ordinary workers to master it and apply it to their own work?

Here's how we teach it (starting with a long narrow strip of paper):

1. First, remember a number: 0.618.
2. Then, we give an example: Suppose the optimal temperature for a certain process lies between 1,000°C and 2,000°C. Of course, we could do a test every degree. After a thousand tests, we'd surely find the best temperature—but it would take a thousand tests.

3. (Take out the strip of paper.) Suppose this strip has a scale marked from 1,000°C to 2,000°C. We perform the first test at 0.618 of the total length. The full range is 1,000 units, so 1,000 × 0.618 = 618. That means the first test is done at 1,618°C. Record the result (fig. 1):

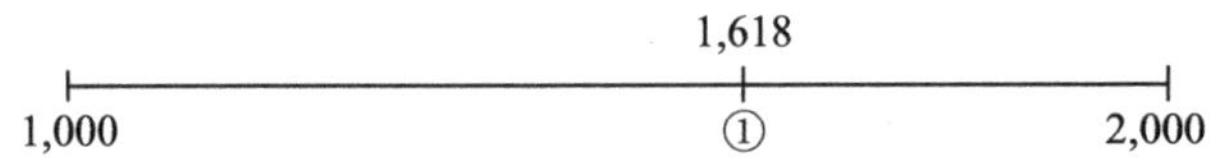

Figure 1

4. Fold the paper strip in half. Perform the second test at the point directly opposite the first one—this is point ② at 1,382°C (fig. 2). Compare the results from the first and second tests. At the worse-performing point (say, Point ①), tear off and discard that part of the strip.

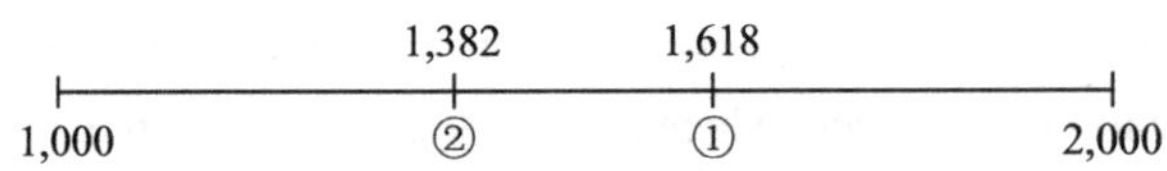

Figure 2

5. Repeat step 4 with the remaining part of the strip, continuing until the optimal point is identified.

This kind of explanation is easy for ordinary workers to understand, and once they understand it, they can use it. Based on the three principles of topic selection we proposed earlier, we selected a number of commonly used optimization and explained them to workers using similar simple and clear language.

For methods that are less easy to popularize, but may be useful in specialized situations, we also conducted in-depth research. For example, the DFP method (Davidon-Fletcher-Powell), proposed in 1962, claims a convergence speed of

$$|x^{(k+1)} - x^*| = o(|x^{(k)} - x^*|)$$

We pointed out that the convergence rate should actually reach

$$|x^{(k+n)} - x^*| = o(|x^{(k)} - x^*|^2)$$

In 1979, while in Western Europe, we learned that W. Burmeister had proven this result in 1973. However, we had already proposed a method in 1968 that achieves this same convergence rate:

$$|x^{(k+n)} - x^*| = o(|x^{(k)} - x^*|^2)$$

This method requires at least half as many tests compared to the DFP method.

■ Fraction Method

At times, real-world conditions do not change in a continuous manner. For example, a lathe may only have a limited number of speed settings. In such cases, the value

$$\frac{\sqrt{5}-1}{2} \approx 0.618$$

might seem hard to apply directly, but continued fractions come into play once again. The continued fraction approximations of $\frac{\sqrt{5}-1}{2}$ are

$$\frac{0}{1}, \frac{1}{1}, \frac{1}{2}, \frac{2}{3}, \frac{3}{5}, \frac{5}{8}, \frac{8}{13}, \ldots, \frac{F_n}{F_{n+1}}, \ldots$$

Here, $\{F_n\}$ represents the Fibonacci numbers, defined by $F_0 = 1$, $F_1 = 1$, and $F_n + F_{n+1} = F_{n+2}$. We used "matchsticks" or small machine parts next to the lathe to explain this method to workers.

For example, suppose a lathe has 12 speed settings: ①②③④⑤⑥⑦⑧⑨⑩⑪⑫. We suggest starting the first test at setting ⑧. Then, using symmetry, the second test is done at setting ⑤. Compare the results to see which is better. If ⑧ is better, discard settings ①–⑤ and keep ⑥⑦⑧⑨⑩⑪⑫. (Otherwise, keep ①②③④⑤⑥⑦) Next, using symmetry again, test setting ⑩. If ⑧ is still better, discard ⑩⑪⑫ and keep ⑥⑦⑧⑨. Continue with symmetry, testing setting ⑦. If ⑦ is better, keep ⑥⑦. Finally, test setting ⑥. If it's better than ⑦, then setting ⑥ is the optimal speed among the twelve, and production can proceed using it.

This method is easy for machine shop workers to master.

■ The Golden Ratio and Numerical Integration

The number

$$\theta = \frac{\sqrt{5}-1}{2}$$

is known as the golden ratio. Not only is it useful in golden section search, it also has a special place in Diophantine approximation, which inspired me to consider the following numerical integration formula:

$$\int_0^1\int_0^1 f(x,y)\,\mathrm{d}x\mathrm{d}y \sim \frac{1}{F_{n+1}}\sum_{t=1}^{F_{n+1}} f\left(\left\{\frac{t}{F_{n+1}}\right\},\left\{\frac{tF_n}{F_{n+1}}\right\}\right)$$

This is a single-sum approximation for a double integral, where $\{\xi\}$ denotes the fractional part of ξ.

How can we generalize this method to multidimensional integrals? The key lies in understanding what $\frac{\sqrt{5}-1}{2}$ represents. It arises from dividing the unit circle into five equal parts. That is, from the equation $x^5 = 1$. That is

$$x^4 + x^2 + x^2 + x + 1 = 0$$

We let

$$y = x + \frac{1}{x}$$

leading to $y^2 + y - 1 = 0$, whose solution is

$$y = \frac{\sqrt{5}-1}{2}$$

or

$$y = 2\cos\frac{2\pi}{5}$$

This is a cyclotomic number. Given that $2\cos\frac{2\pi}{5}$ is useful, what about other numbers of the form

$$2\cos\frac{2\pi l}{p},\ 1\le l\le\frac{p-1}{2}=s$$

where p is an odd prime? Could they help us perform high-dimensional numerical integration?

Minkowski's theorem has long proven that there exist $x_1, \ldots, x_{s-1}$ and y such that

$$\left|2\cos\frac{2\pi l}{p}-\frac{x_l}{y}\right|\le\frac{s-1}{sy^{s/s-1}}$$

But Minkowski's proof is only existential. However, for the cyclotomic field

$$R\left(2\cos\frac{2\pi}{p}\right)$$

because we can explicitly express a system of independent units, we can effectively construct such $x_1, \ldots, x_{s-1}$ and y. Therefore, we can use the sequence

$$\left(\left\{\frac{t}{y}\right\},\left\{\frac{tx_1}{y}\right\},\ldots,\left\{\frac{tx_{s-1}}{y}\right\}\right), t=1,\ldots,y$$

to replace

$$\left(\left\{\frac{t}{F_{n+1}}\right\},\left\{\frac{tF_n}{F_{n+1}}\right\}\right), t=1,\ldots,F_{n+1}$$

This approach is not only suitable for numerical integration, but also for any situation where random numbers are used.

■ Overall Planning Method

Educational reform should not only broaden students' knowledge but also serve the development of societal production. The optimization introduced above are not only easy to popularize but also practical tools for improving production processes.

For example, quality control usually aims to prevent defective products from leaving the factory, thus maintaining the reputation of a company's product quality. However, instead of dealing with defects after they occur, it's better to first apply optimization to determine the best production conditions and reduce the defect rate from the start. Once defects are minimized, quality control becomes much easier and more effective.

In addition to managing production techniques, there is also the issue of managing production organization. The mathematical methods used to address such organizational challenges fall under what we call the method of overall planning, or operations research.

There are many effective methods within this field that are suitable for popularization. Here are a few examples:

CPM. When we began popularizing this method, we made it easier to understand by focusing on a simple goal: minimizing the project duration. But once people grasped the concept, they could tackle more complex issues such as minimizing investment or balancing labor and resource allocation. Everyone now knows what CPM is, so instead of explaining the theory, I will describe how we implement it in practice.

Our primary principle is to base the approach on real-world projects, enabling technicians or workers to learn this method. The steps are as follows:

1. Investigation. Items for investigation are a list of all relevant activities, the connective relations between activities, and the duration of each activity. To do this well, it is essential to rely on frontline workers and technicians, as their estimates are often more accurate and realistic than those made by higher-level technical staff.
2. Critical path diagram. Based on this information, we guide everyone to draw preliminary diagrams, then teach them how to identify the critical path. After that, the team discusses, shares ideas, works together to shorten the project duration, sets the schedule, and finally draws the CP diagram.
3. Heed to the transformation of contradictions. During the course of a project, tasks are often completed ahead of or behind schedule, so the critical path is not fixed. We must continuously monitor these changes and give timely instructions to the relevant work sections.
4. Summing-up. At the completion of a project, the actual stage-by-stage progress is recorded in the form of CP diagrams so that this information may be utilized for future projects of a similar type.

From our experience, this method works well for small-scale operations and better for large-scale projects. It can start from the grassroots level, with each work section gradually contributing to the full project CP diagram. Alternatively, one can take a top-down approach by first drafting a broad, high-level plan, then having each unit develop its own CP diagram based on that framework, followed by collective discussion and revision.

Sequencing analysis. Suppose there are several tasks (or projects), each with an estimated duration (possibly obtained using CPM). If the order of execution is flexible, how should the sequence be arranged to minimize total waiting time?

To solve this, we start with a mathematical principle:

Given two lists of non-negative numbers: $a_1, a_2, \ldots, a_n$ and $b_1, b_2, \ldots, b_n$, how should the numbers be ordered to minimize or maximize the sum:

$$\sum_{i=1}^{n} a_i b_i$$

The answer is that the sum is maximized when the sequences are in the same order, and minimized when they are in reverse order. This can be proven easily, even from the simplest case: If $a_1 \le a_2$ and $b_1 \le b_2$, then $a_1b_1 + a_2b_2 \ge a_1b_2 + a_2b_1$, which is equivalent to $(a_2 - a_1)(b_2 - b_1) \ge 0$. This tells us that correcting any mismatched pair of orders toward same-order alignment increases the total sum.

To put it more plainly: Imagine a tap and n buckets with capacities $a_1, a_2, \ldots, a_n$. In what order should the buckets be filled to minimize total waiting time? The first bucket takes a_1 time to fill, the second takes $a_1 + a_2$, and so on. Total waiting time becomes

$$a_1 + (a_1 + a_2) + \cdots + (a_1 + a_2 + \cdots + a_n) = na_1 + (n-1)a_2 + \cdots + 2a_{n-1} + a_n$$

which is minimized when the sequence a is arranged in reverse order to b, that is, when $b_1 = n, b_2 = n-1, \ldots, b_n = 1$. That is, when the smallest buckets are filled first:

$$a_1 \le a_2 \le \ldots \le a_n$$

If there are s water taps, and the sequence of bucket capacities at the first tap is $a_1^{(1)}, \ldots, a_m^{(1)}$, at the second bucket $a_1^{(2)}, \ldots, a_m^{(2)}, \ldots$, and so on, then the total waiting time is

$$\sum_{j=1}^{s}\left(ma_1^{(j)}+(m-1)a_2^{(j)}+\ldots+a_m^{(j)}\right)$$

(We do not rule out the possibility that some $a_t^{(j)}=0$.) Define

$$b_1=b_2=\cdots=b_s=m, b_{s+1}=b_{s+2}=\cdots=b_{2s}=m-1,\cdots\cdots$$

Then we arrive at the conclusion: It is still best to "fill the smaller buckets first."

The two basic examples above provide everyone with a preliminary understanding of how to manage multiple projects with overall planning. Next, we introduce methods that are practically useful for those responsible for organizational management.

Another method that can be widely taught is the graph-based operations method for transportation and allocation. Suppose there are n wheat production sites a_1, …, a_n, each producing A_1, …, A_n tons of wheat, and m consumption points, each requiring B_1, …, B_m) tons. The goal is to minimize the total ton-kilometers (i.e., the product of transported weight and distance). This problem can certainly be addressed using linear programming, but we often employ a simpler graph-based method. The key principle of this method is to use a transportation graph to eliminate counterflows and detours.

▪ The Use of Statistics

▸ Empirical Formulas and the Importance of Mathematical Insight

Sometimes an empirical formula is derived from a set of statistical data. The meaning of such a formula can often be recognized by scientists with good common sense and some mathematical training. Let me give an example. R. C. Bose, an Indian mathematical statistician, obtained the empirical formula

$$A=\frac{\text{Length}\times\text{Width}}{1.2}$$

for the area A of a leaf of rice grain, from a large sample of such leaves in India. I had no reason to question its reliability. However, some agriculturists in China applied this same formula to the harvest on their rice farm. On seeing the shape of the leaves on their farm, I pointed out to them that this formula could not be suitable for their leaves. They took some samples of their leaves and discovered that the formula tended to overestimate their areas. After their expression of surprise, I gave them a simple explanation by drawing the following picture (fig. 3), where the shaded area represents the area of a leaf.

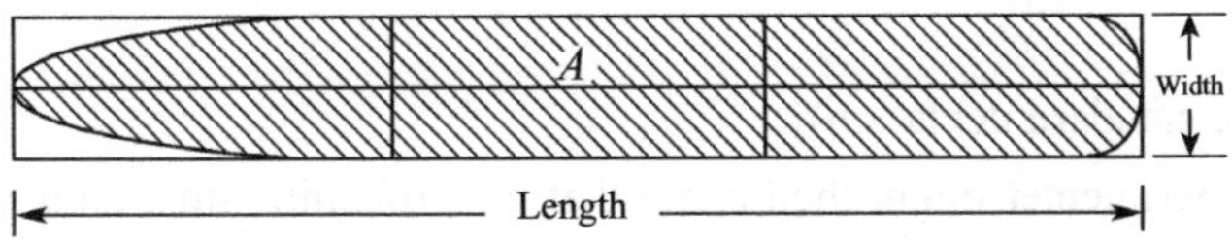

Figure 3

In this case, the ratio of the area of the rectangle to A is 6/5, i.e., 1.2. In contrast to this, a typical leaf in their farm has a more elongated tip, and I drew another picture (fig. 4). At this point, the ratio of the rectangular area to area A is clearly greater than 1.2. It is easy to explain why using Bose's formula would result in an overestimation of the area of their rice leaves.

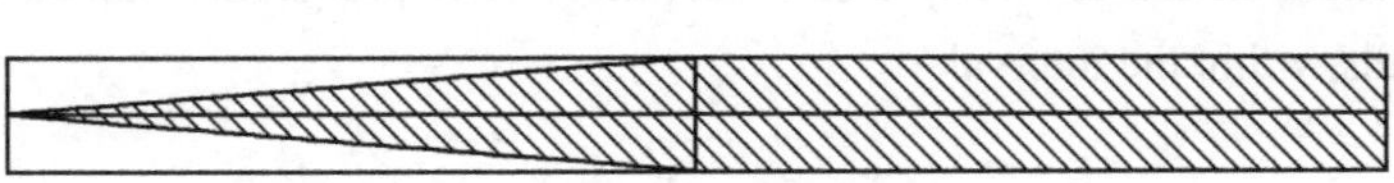

Figure 4

We all learned the lesson of the importance of seeking mathematical insight beneath an empirical formula.

A Simple Statistical Method

In the experimental sciences we frequently use statistical methods. Needless to say, their importance must not be denied. However, it is my view that some methods tend to be too sophisticated, and can easily be misused. I give a few examples first:

1. Let $x_1, \ldots, x_{20}$ denote twenty observations from independent repetitions of an experiment. Let

$$\bar{x} = \frac{(x_1 + \ldots + x_{20})}{20}$$

(the sample mean) and

$$s = \sqrt{\frac{\sum_{i=1}^{20}(x_i - \bar{x})^2}{19}}$$

(the sample standard deviation).

The experimenter might then claim that the probability that an observed value falls within the interval

$$\left(\bar{x} - 1.73\frac{s}{\sqrt{20}}, \bar{x} + 1.73\frac{s}{\sqrt{20}}\right)$$

is 0.9. Such a sophisticated method does not seem to be easily understood by an ordinary worker in China. Besides, the underlying Gaussian assumption might not be valid in his case!

In practical terms, I tend to prefer the following approach: Let $x_{(1)} \leq x_{(2)} \leq \ldots \leq x_{(20)}$ denote an ordered sample of observations. The probability that an observed value falls within the interval

$$\left(\frac{x_{(1)} + x_{(2)}}{2}, \frac{x_{(19)} + x_{(20)}}{2}\right)$$

might be greater than 18/20 = 90%.

2. Suppose that it is required to test which of two methods of production is better on the basis of five observations from each method. Let $\{a_1, \ldots, a_5\}$ and $\{b_1, \ldots, b_5\}$ denote the samples of the first and second methods respectively. It may be tempting simply to compare the means of the two samples by referring to the usual student's t distribution. Underlying such a sophisticated method is a number of assumptions, such as normality, equal variance and independence, which are not easily understood by an ordinary factory worker.

It seems to me that a more robust and simple method based on the ordered samples $a_{(1)} > a_{(2)} > \ldots > a_{(5)}$ and $b_{(1)} > \ldots > b_{(5)}$ may be more suitable for popularization in China. For example, if the mixed ordered sample is, say, either

$$a_{(1)} > a_{(2)} > a_{(3)} > a_{(4)} > b_{(1)} > a_{(5)} > b_{(2)} > b_{(3)} > b_{(4)} > b_{(5)}$$

or

$$a_{(1)} > a_{(2)} > a_{(3)} > a_{(4)} > a_{(5)} > b_{(1)} > b_{(2)} > b_{(3)} > b_{(4)} > b_{(5)}$$

I would usually demonstrate the former by holding my two hands up with the two thumbs crossing each other (fig. 5).

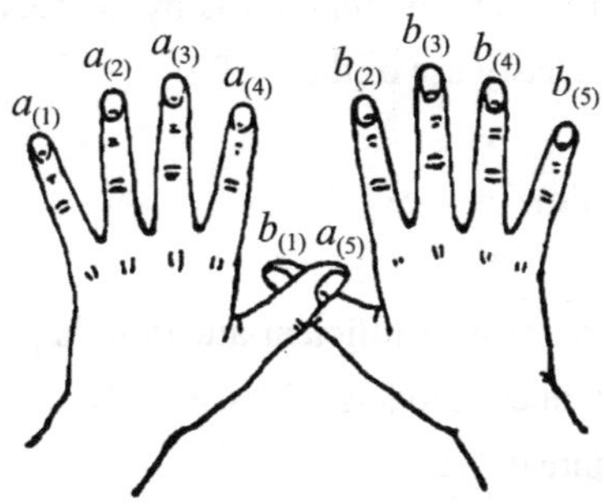

Figure 5

An ordinary factory worker can accept it as being sensible to reject the hypothesis that the two methods are equally good. Furthermore, there are 25 possible comparison outcomes between the two sets of samples ($5 \times 5 = 25$). Except for the case where $b_{(1)} > a_{(5)}$, in all other cases, "*a*" is greater than "*b*." Therefore, the statement "the first production method is better than the second" has a 24/25, or 96%, probability of being true.

PERT

Consider the so-called Program Evaluation Review Technique (PERT). Suppose that there are *N* activities in a network representing a project and there are three basic parameters describing the duration of the *i*-th activity. Thus, let a_j, b_j and c_j denote the "optimistic duration," the "most probable duration" and the "pessimistic duration" respectively. The duration of the *i*-th activity is then usually assumed to follow a beta distribution over (a_j, c_j) and to have mean duration m_j where

$$m_j = \frac{(a_j + 4b_j + c_j)}{6}$$

and variance

$$\frac{(b_i - a_i)^2}{36}$$

It is often argued that the probability distribution of the total duration of the whole project may be approximated by a Gaussian distribution, which is disputable, presumably by appealing to the Central Limit Theorem (CLT). Even ignoring the fact that the assumption of beta distribution is by no means indisputable, I would question the wisdom of a hasty use of the CLT.

▸ Experimental Design

Up to now, it seems to me that insufficient attention is paid to non-linear designs. Past preoccupation with linear models still seems to mask the important fact that these models are often unrealistic.

We must continually refine our models to bring them closer to reality. At the same time, we recognize that no model is ever a perfect replica of the real world, and it's unrealistic to expect one to fully realistic.

▸ Types of Distributions

It has been suggested that the Pearson's Type III distribution be used to model the distribution of the waiting time for an "exceptionally big" flood. I would question the validity and wisdom of this approach in view of the scarcity of data which is inherent in this problem. It is even more unwise to expect that a sensible forecast for the next big flood may be obtainable from such a fitted distribution!

■ Mathematical Model

▸ Generalized Inverse of Matrices

Consider a general regression model of y on $x_1, \ldots, x_p$, $y = f(x_1, \ldots, x_p) + e$, where e denoting the random "error" term. Let $y^{(i)}$ denote the observed value of y at $x_1^{(i)}$, $x_2^{(i)}, \ldots, x_p^{(i)}$. If f is assumed linear and there are $n (> p)$ observations, then

$$y^{(i)} = \sum_{j=1}^{p} \theta_j x_j^{(i)} + e^{(i)}, i = 1, \ldots, n \tag{1}$$

A general method for estimating $\theta_1, \ldots, \theta_p$ is by minimizing

$$\sum_{i=1}^{n} [e^{(i)}]^2$$

with respect to θ_j. That is, it is required to minimize with respect to $\underset{\sim}{\theta}$,

$$Q(\underset{\sim}{\theta}) = (\underset{\sim}{y} - \underset{\sim}{M}\underset{\sim}{\theta})'(\underset{\sim}{y} - \underset{\sim}{M}\underset{\sim}{\theta})$$

where

$$\underset{\sim}{y} = [y^{(1)}, \ldots, y^{(n)}]'$$

$$\underset{\sim}{M} = \begin{pmatrix} x_1^{(1)} & x_2^{(1)} & \cdots & x_p^{(1)} \\ \vdots & \vdots & & \vdots \\ x_1^{(n)} & x_2^{(n)} & \cdots & x_p^{(n)} \end{pmatrix}$$

$$\underset{\sim}{\theta} = (\theta_1, \ldots, \theta_p)'$$

Now, for simplicity it may be assumed that M is of rank p. Then

$$Q(\underset{\sim}{\theta}) = [\underset{\sim}{\theta} - (\underset{\sim}{M}'\underset{\sim}{M})^{-1}\underset{\sim}{M}'\underset{\sim}{y}]'\underset{\sim}{M}'\underset{\sim}{M}[\underset{\sim}{\theta} - (\underset{\sim}{M}'\underset{\sim}{M})^{-1}\underset{\sim}{M}'\underset{\sim}{y}] + \underset{\sim}{y}'[I - \underset{\sim}{M}(\underset{\sim}{M}'\underset{\sim}{M})^{-1}\underset{\sim}{M}']\underset{\sim}{y}$$

$$= S_1 + S_2$$

Because $S_1 \geq 0$, therefore $Q(\underset{\sim}{\theta})$ is minimized on setting

$$\underset{\sim}{\theta} = (\underset{\sim}{M}'\underset{\sim}{M})^{-1}\underset{\sim}{M}'\underset{\sim}{y} \tag{2}$$

which is sometimes regarded as a generalized solution of the equation

$$\underset{\sim}{y} = \underset{\sim}{M}\underset{\sim}{\theta} \tag{3}$$

For this reason $(\underset{\sim}{M}'\underset{\sim}{M})^{-1}\underset{\sim}{M}'$ is called a generalized inverse of $\underset{\sim}{M}$.

Of course, if the true model is adequately given by equation (1), then solution (2) is valid. Otherwise, if by substituting solution (2) in equation (3), substantial discrepancies are observed between the "observed" values and the "predicted" values of y, then the assumption of linearity should be abandoned.

The use of generalized inverse is only one of the many examples of solving an essentially non-linear problem by posing it in a linear form. Other examples are linear programming and orthogonal design, to which the above remarks still apply.

▸ Matrices with Non-Negative Elements

Since many economic variables are typically non-negative, the theory of matrices with non-negative elements is rather suitable for the analysis of economic relation. I believe that this theory has much to offer for the modeling of the Chinese economy.

■ Concluding Remarks

If I were asked to say in a few words what I have learned in these last fifteen years of popularizing mathematical methods, I would without hesitation say that they have enabled me to appreciate the importance of the dictum (fig. 6):

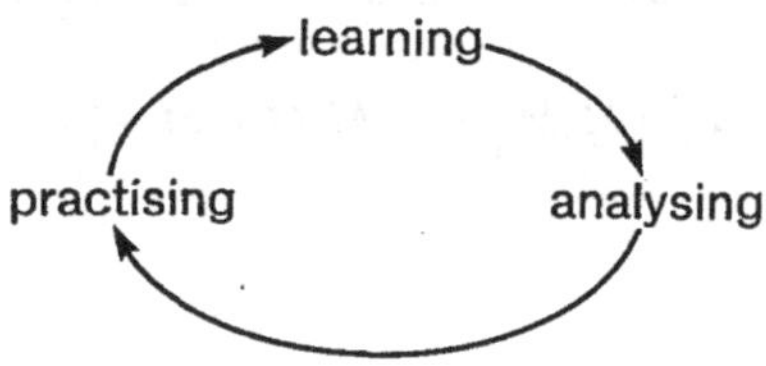

Figure 6

Unforgettable Memories*

■ I

It was a late autumn night in the early 1970s. Two people were deep in conversation in a guest room. Sitting across from each other at a square table, they gradually leaned closer, eventually talking at one corner of the table as their discussion deepened. The grandchildren, sensing the atmosphere, had gone to bed early.

"Could you talk about your experiences over the past few years—bringing mathematics to practical use in the Wumeng Mountains, along the Dadu River?" The guest's sincerity, concern, support, and thoughtful attitude moved me deeply. I felt I had a thousand things to say, but didn't know where to begin.

For me, this was something I could never have imagined—even in my dreams. The study and the classroom had always been my world. Especially for someone like me, who had pursued a single subject in isolation and was largely self-taught, mathematics was the only area I had managed to grasp. As for other fields of knowledge, I had almost nothing. I envied those who had received a formal education, who had at least a basic foundation in science, nature, and the world.

So for me, taking this step was no easy feat! But in reality, it wasn't as hard as I had imagined. This was something everyone cared about. Once I found a common language and shared goals with the people, everything became easier. When united in purpose, even the hardest tasks become manageable. And besides, weren't there people like the one sitting beside me now—talking heart to heart, showing care, offering encouragement, support, and guidance?

Weren't they the same people who had once crossed snowy mountains and marshy grasslands, and liberated China with only millet and rifles? Could we really

---

* First published in issue 31 of *Outlook Weekly* in 1984.

be daunted by this small challenge? Would the fleeting fame of academic authority get in the way of serving the people? Everyone has different abilities, but no good deed is too small. Every contribution counts.

Of course, these weren't the exact words I used that night, but they reflect the spirit and principles that underpinned what I was sharing. The guest listened quietly, occasionally asking questions. Our minds resonated. The more we talked, the deeper the conversation grew. Then he suddenly asked, "What are your plans for your future work?" Because I had been thinking about this for a long time, I answered without hesitation: "Comprehensive planning, broad optimization, integrated transportation, and grand balance." He paused in silence for a moment—I could tell he was giving it serious thought. "I agree with your direction and with your idea of finding research topics in real-life practice," he said. "But could we perhaps revise those last two words a bit? Balance is always temporary and relative. All things are in motion. So could we rethink those final words?" His attitude was kind and humble, and he spoke in a consultative tone. He guided me to reflect from the perspective of the principles of human social development, inspiring me to revise my own thinking. That was far more effective than any long-winded lecture or a blunt criticism about being "rigid-minded"!

At the time, I can't see the wood for the trees. I was fixated on the immediate losses and difficulties our country was facing due to imbalance. That current reality clouded my view, making me forget that continuous development is the true path of social progress. So, I wondered—might it be better to replace "grand balance" with "strategic development"? He nodded in agreement. Later, this change was tested repeatedly in real-world practice and explored more deeply through mathematical theory. It turned out that in a planned economy, "strategic development" was not only more accurate but also far more proactive than "grand balance."

Late into the night, the autumn chill flowed like water. The respected guest had left behind theoretical insights and directional guidance, allowing me to vaguely grasp the relationship between science, society, and philosophy. I walked him out, yet could not keep up with his calm and swift steps, finally stopping to watch him leave. The western wind blew fiercely, frost gathered thick on the ground, and without a hat or scarf, his figure disappeared into the misty moonlight.

Although by that time Lin Biao and his ilk had already self-destructed, the country was still caught in the whirlpool of the "Cultural Revolution." His words filled me with confidence and resolved me with determination: I had to immerse myself in real-world production to seek out problems, and to approach them from the perspective of development.

■ II

Autumn, 1982.

A hospital room. A hospital bed. A monitoring device emitted rhythmic wave signals. Two nurses concentrated intensely, recording data non-stop. Lying on the bed was a patient—me. I heard the attending physician had already reported to the leadership that the situation was critical. It seemed like another myocardial infarction. A "No Visitors" sign hung on the door, and the nurses gently explained to those who came to see me that the patient needed complete rest and must not be disturbed or made aware of any visitors.

The patient lay quietly, wordless, as if in complete repose—but from time to time, my eyes would open and close, flickering with emotions both joyful and anxious. It was clear that despite the calm appearance, my mind was churning beneath the surface. I was deep in thought, exploring possibilities.

Was this truly like looking for a needle in a haystack? In some ways, yes. But the haystack is finite—whereas the realm of thought is boundless. That "needle," that single thread of insight, had been lost for more than a decade. I had already spent two years actively trying to find it again. I once thought that it would be easy to retrace old steps—familiar paths, skills honed through experience. Surely, I could quickly recapture the theoretical ideas I had previously developed on how to apply mathematical methods more effectively to a planned economy.

But things didn't go as expected. Around the time of the 12th Party Congress, I had the feeling that the handwritten manuscripts I had toiled over in the 1960s—developing these very theories—might finally find use. Back then, I had written a line: "Too little, too late—looking for a solution that should have been prepared well in advance." I had intended to prepare in advance, to make some contribution to the country. Now it seemed that the early preparations were finally needed—what a joyful realization that was! I wished I could write it all out in one go, present it as a gift to the Party. But things did not go as planned. It felt like trying to find something lost in endless smoke and sea—completely beyond reach.

Looking back to the late 1950s and early 1960s, I had already been considering the application of mathematical methods to planned economics. The conceptual blueprint had taken early shape. This idea needed to be gradually revised, supplemented, and elevated through real-world practice—starting from the workshop, from production units, laying a strong foundation before raising pillars and beams.

The early stages were just what I had mentioned before: "comprehensive planning, broad optimization, integrated transportation." On this foundation, the final step would be "strategic development"—structurally, that was the roof of the building.

During the "Cultural Revolution," from the very beginning, some people with "a keen eye for sabotage" had targeted this goal, ignoring the Party's directive of integrating theory with practice. Yet a few upright leaders, even at the risk of being labeled as adherents of the "theory of productive forces," supported us in conducting pilot programs. Our development expanded from workshops to factories, from enterprises to entire provinces and cities. This allowed us to engage with hundreds of thousands of industrial leaders, engineers, technicians, and working people. Their support and experience taught us much and gave us a solid foundation for those first six words. But in that chaotic era, how could we even begin to talk about "strategic development"? It was in the mid of the chaos that we fought for time and gradually prepared the conditions needed for "strategic development." But three months passed—and I still hadn't made any progress. The theories I had once developed had vanished completely from memory. Having spent my life in research, I always knew that discovery and invention were difficult—but I had no idea how hard it would be to retrieve a thought once it had been forgotten.

■ III

The attending doctor quietly came in to check on me and assess my condition.

She gently asked, "Have you been overworked lately? The newspaper reported on the success of your work at the Lianghuai coal mines, and the leadership even praised your team. Have you been exhausting yourself?"

No, the work at Lianghuai wasn't what wore me out. Among the sixty-plus people who worked together with me, along with over a hundred others from our own unit, everyone else was busier than I was—working around the clock without rest. Everyone was protective of me, trying their best to reduce my workload and let me rest more, shielding me from the trivial matters. Just like every time I've gone to various provinces and cities, our team was always united. Even though the members came from five different ministries and seven academic societies, all the experts worked with great care and dedication. In such a cooperative, achievement-oriented group—and with three months having passed since then—this was unlikely to be the root cause of my illness. The doctor nodded and left.

When she returned later, her wording had completely changed. She said, "I've realized you're the kind of person whose mind never rests. Instead of trying to 'order' you to stop thinking—which is impossible—it's probably better to let you focus wholeheartedly on the problem you consider most important. Even though I don't understand the specifics, I know you're thinking about something that would benefit the people. But don't underestimate the fact that you're still a seriously ill patient. We're monitoring your condition closely with the equipment, and we need you to follow our medical advice. Even if you start to feel better, we're not planning to take down the 'No Visitors' sign on your door until you've calmly found the line of thought you've been searching for."

This wise and thoughtful approach relieved my mental burden. In truth, my work had never actually stopped. By the time I was discharged three months later, the entire line of reasoning had finally come together. But one key theorem's proof continued to trouble me—I couldn't recall the simple, clear proof I'd come up with back in the 1960s. Determined, I pushed through, and eventually found a more complicated proof. At last, the framework was basically complete.

Yet it seemed that this proof wasn't in the same style as my past work—it left me feeling slightly unsatisfied. In September 1983, I wrote seven abstracts and sent them to *Science Bulletin*.

When I left the hospital, full of joy, a poem I had written after being discharged from my first heart attack echoed once more in my mind:

The steeds of Hulunbuir,
The eagles of Mount Everest,
Gallop across the plains with hearts set on distant horizons,
Soar through the skies with ambition that reaches the clouds,
All for the people.

A warrior facing death in battle cares little for wounds,
Charging at ancient evils for the sake of a new era,
Willing to give all, without regret.

■ IV

Once the core of my original theory had become clear, I was fortunate enough to receive an opportunity to go abroad. After more than twenty years, I wondered—had those ideas from the 1960s already become obsolete relics? Now, I had the chance to exchange views with scholars from around the world.

I've always disagreed with the old Chinese saying, "Don't show off your skills before an expert." On the contrary, I believe the true spirit lies in "Do show off your skills before an expert." And so, despite feeling anxious and unprepared, I decided to take the risk. I knew almost nothing about economics and yet I was likely about to embarrass myself. Still, I reminded myself: Exposing one's inadequacy offers more chances for learning than hiding one's shortcomings. So, I pushed past my nerves and forged ahead.

This trip abroad allowed me to compare the fruits of my quiet, self-contained labor against the most renowned schools of economic thought. To my great relief, I found that this approach had not yet been explored by foreign scholars. At the same time, my new proof was finally completed—surprisingly, it once again echoed the style of my original "direct method" from all those years ago.

Just as I was preparing to pen a letter to the esteemed leader who had encouraged me to record my recollections, Xinhua News Agency and *People's Daily* published a letter from Comrade Hu Yaobang to me, reprinted from *Liaowang Weekly*. Looking back now, I realize that it took me a full two years to fulfill even one of his instructions. I was fortunate: Thanks to the help of doctors, friends, and the chance to interact with many scientists abroad, I was able to make progress. Otherwise, even if I had finished the work, I might have doubted whether foreign scholars had already done it. I am sincerely delighted that the ideas I conceived in the 1960s, despite the passage of two decades, had not fallen behind the times. That, I can only call a blessing from above. Looking at it objectively, perhaps this result was destined to emerge around this time, and I was merely fortunate enough to arrive there a step earlier.

Returning once more to the eight words that shaped my early thinking: The first six—"comprehensive planning, broad optimization, integrated transportation"—are methods rooted in not seeing the wood for the trees. These are techniques I, along with my colleagues, had worked to refine and promote for over twenty years: systems like overall planning, optimization, linear programming, and the distilled essence of other classical methods. But the last two—"strategic development"—represent an entirely different perspective, one that allows you to see the wood. When combined, the two approaches form a complete framework for managing a planned economy—and, as some foreign experts have suggested, they may even offer insights for free-market economies.

This whole journey has taught me a deep lesson: Any meaningful work must be rooted in practice, beginning at the grassroots and expanding step by step, with each footprint firm and real. At the same time, a solid foundation in theory is the only way to transform perceptual knowledge into structured understanding.

I even wrote a lighthearted poem to sum up this lesson—perhaps it will serve as a caution for beginners tempted to skip the hard work at the roots and leap straight for the treetops:

On the Difficulty of Using Talent

Du Fu once wrote in his poem "The Ancient Cypress,"
defending the great tree,
Let me rebut his words,
For this tree, at least, was lucky enough to grow in a protected courtyard.
When still a sapling, it risked the hooves and teeth of cattle and sheep;
When grown, it faced the axes and saws of woodcutters.
How could it ever reach two thousand feet skyward,
If not for the deep affection left behind by the Prime Minister?
The old saying, "Big wood is hard to put to use," always felt incomplete,
Big timber can always be cut down and divided, shaped into countless tools.
Small beams can hold up great halls,
But great logs, if used unwisely, can mislead a nation or ruin a life.
Never dismiss small tasks as beneath you,

For small deeds are the roots of great virtue.
Don't let pride in your own "greatness,"
Make you shallow and boastful, wasting your potential.

Though this is true for the individual,
For the nation, overall planning must reign.
Let science and division of labor make full use of talent,
And let foresight and vision nurture the capable and the wise.

■ V

Self-awareness is a rare and precious thing. The old saying goes, "Love the house and its crow." The house is the great edifice of the motherland, and as for me—I am nothing more than a crow perched quietly in one corner of its roof. This is no false humility; I say this from the heart.

I recall an instance that left a deep impression on me: When Professor Frank Press, then President of the US National Academy of Sciences, broke precedent and read the commendation for my election as a foreign academician, after presenting my scientific accomplishments, he added a thoughtful remark: "He is self-taught, yet he has taught millions." At that moment, the hall erupted into warm applause. That applause reminded me that in the past, I had been the kind of person who enjoyed burrowing deep into thought and climbing high in search of knowledge, content to hide away in the ivory tower. Without the Party's guidance, I would never have had the chance to work side by side with millions of ordinary people, or to immerse myself in real production and practice. Such opportunities were rare for scientists of earlier generations.

I am deeply grateful for the letters of congratulations from so many people. Comrade Fang Yi also wrote to encourage me: "You are the first Chinese scientist in the 120-year history of the US National Academy of Sciences to be awarded this honor." But in truth, this was the voice of the Party, urging me to lay one more brick, to place one more tile atop the Tower of Knowledge. Brick by brick, effort by effort.

And beyond this, patriotic compatriots from overseas, from Taiwan, Hong Kong, and Macau, all vied to take photos with me as keepsakes. I remember an elderly overseas Chinese, his hands strong as he held out a pot of American ginseng, his eyes brimming with tears: "Your honor is the pride of us all. This ginseng is our wish for your long life, so that you may continue to serve the motherland."

Even the criticisms we've received—pointing out where our work still falls short—have been of great help. These critiques were born from hope: that we would go from good to even better. Otherwise, why would they bother? For those who sympathize with our efforts in harsh conditions, our gratitude is beyond words—and it drives us to work a hundred times harder. But even if that sympathy weren't there, the pursuit of science is a pursuit of truth. If the truth eludes us, we must keep striving until we match, and surpass, those who, with better resources, have climbed ahead of us. Only through unity can we complete the Tower of Babel.

And in the end, there is only one thing left to say: "When drinking water, remember the source." This is dedicated to the person who inspired us to step into real production, to approach work with the heart of a true master—that plainspoken, clear-sighted man: Comrade Hu Yaobang.

Pouring Heart and Soul into Cultivating Talent*: Reflections Inspired by Reading *Selected Works of Zhou Enlai* (Volume II)

As I lay on my sickbed, my hands trembling and my breath short, I received *Selected Works of Zhou Enlai* (volume II). Despite my frail health, I couldn't resist reading it hungrily, page after page. The deeper I read, the more it stirred my memories of Premier Zhou—memories so vivid they kept me wide awake, deep in thought, until late into the night. I felt compelled to write down my reflections. I drafted two versions, but my handwriting had become so shaky that even the colleague who tried to copy them couldn't recognize some of the words—neither could I. In the end, I had to rely on someone to write down what I dictated from my heart, though even so, I felt my words could never fully express the depth of my admiration and longing. After all, how could mere ink and paper capture the profound love and respect I hold for Premier Zhou?

■ From Theory to Practice

Let me begin with the connection between theory and practice. I was a self-taught scholar—a solitary learner—whose knowledge extended little beyond certain

---

* First published in the *Economic Daily* on January 7, 1985.

branches of mathematics. That was the true state of my knowledge when I returned from the United States in 1950.

Shortly after my return, I studied Premier Zhou's speech at the National Conference on Higher Education. Even now, some of his words remain etched in my memory: "Learning theory requires repeated practice in order to grasp it more accurately and understand it more deeply. To neglect practice or to pit theory against practice is simply wrong." "Whether one is a generalist or a specialist, the connection between theory and practice is essential." (*Selected Works of Zhou Enlai*, vol. II, 18)

At the time, his words directly addressed both my mindset and the reality I faced, prompting me to rethink the habitual way I approached problems. I had been tasked with helping to rebuild the Institute of Mathematics, which, before the liberation, had moved to Taiwan and only covered a few branches of pure theoretical mathematics.

When we re-established the institute, with the help of colleagues, we not only restored the core theoretical disciplines but also added emerging interdisciplinary fields—such as mathematical logic, mechanics, mathematical physics, computing, and computer science. Our vision then was to create an interconnected academic framework, each discipline linked to the next like rings in a chain. But whether this chain truly connected, whether it ultimately served the national economy, was something we gave little thought to at the time. Premier Zhou said, "Our universities must study theory, but it must be theory that has been tested by practice, and the goal must be to use this theory to further guide practice and better serve the people." (Ibid.) In hindsight, our thinking back then was still far from living up to this ideal—the genuine unity of theory and practice that Premier Zhou emphasized.

In 1958, during the Great Leap Forward, the imbalance in the national economy became starkly apparent, shaking me to the core. As a patriotic scientist, how could I remain indifferent to this reality? But as a mathematician, I asked myself: What could I do? Could mathematical methods be applied to the planning and adjustment of the national economy?

I began with site visits and field studies, then moved on to hands-on experiments at the grassroots level. With these practical foundations, I worked to develop clear and accessible methods. Eventually, I completed one "theory" (the mathematical theory of large-scale optimization for planned economies) and two "methods"—overall planning and optimization. I even wrote popularized explanations for both, so they could be easily understood, learned, and applied by the masses. This was my humble attempt at applying the connection of theory to practice in mathematics. The most essential quality for any scientist is a commitment to truth and pragmatism. The two methods I devised were not born from abstraction but distilled from dozens of approaches—pared down, refined, and repeatedly compared until only the most practical and effective remained. Without the guidance of Premier Zhou's principle of "doing real work, achieving real results," these two methods could never have been created. For more than two decades, these methods have been adopted and applied by countless grassroots communities, their success a small but heartfelt tribute to the memory of our beloved Premier.

■ The Care, Encouragement, and Support He Gave Me

I will never forget the deep care, encouragement, and unwavering support Premier Zhou Enlai gave me over the decades. What I remember most fondly—and with the deepest gratitude—was his care during the most difficult period of the "Cultural Revolution," particularly in 1970, a year of immense hardship. At that time, Premier Zhou himself was facing grave adversity and was burdened with countless state affairs. Yet despite the dangers to his own safety and his declining health, he still made the effort to protect me with meticulous care, thoughtfully arrange my living conditions, and show concern for our work in applying mathematical methods to economic development.

I still vividly remember one Saturday evening, when two officials from the State Council came to deliver Premier Zhou's written instructions, issued on March 4, 1970:

> First and foremost, Hua Luogeng must be protected to prevent any harm from ill-intentioned people.
>
> Second, a thorough investigation must be conducted into the theft of his manuscripts, and every effort should be made to solve the case.
>
> Third, Institute of Mathematics of Chinese Academy of Sciences, which has sealed his personal collections, must have Xiyao carefully check for any signs of theft and consider returning them to him if proper safeguards are in place.

> Fourth, given Hua's health, his lifestyle is no longer suitable for relocation to the "May Seventh Cadre School" with the University of Science and Technology or for assignment to a distant location. The best course would be for him to remain in Beijing, under the status of a member of the Standing Committee of the National People's Congress, and continue to test the mathematical overall planning he has advocated.
>
> Once the three of you have completed these arrangements, report back to me.

(*Selected Works of Zhou Enlai*, vol. II, 455)

When I heard this directive, I was so overwhelmed with emotion that words failed me. The only way I could express my deep respect for Premier Zhou was by reporting on the progress we had made in serving the national economy through mathematical research. After hearing my report, the two officials told me that although the next day was Sunday, they hoped I could introduce overall planning at a meeting of department heads at the State Council on Monday morning. Deeply honored, I presented the method that Monday. Afterward, one of the attendees remarked, "You've also developed Optimization—why not share that with us as well?" And so, for the first time, I formally introduced Optimization to the leadership.

That same summer, with Premier Zhou's approval, the three of us—my students and I—went to Shanghai to promote and pilot overall planning. Our experiments on optimization quickly yielded a series of highly effective results. But as our work attracted attention, it also drew the notice of Wang Hongwen's circle, who began to "pay extra heed" to us. I was moved from the Peace Hotel to the heavily guarded Yan'an Hotel, and they cut off all contact between us and the public, making it impossible for us to continue our work. When I mentioned this to Comrade Peng Chong in a phone call, he immediately invited us to continue our work in Jiangsu Province. But even this was blocked by the same forces. It was only after Comrade Liu Xiyao personally visited the project site in Shanghai, confirmed the value of our work, and insisted that due to my health I should return to Beijing for treatment, that we were finally able to leave safely.

Back in Beijing, we resumed the pilot testing and practical application of the "Two Methods," steadily expanding across industries and achieving increasingly remarkable results. The State Council, upon hearing of this success, convened another meeting at the State Council Assembly Hall, where department leaders were invited to listen to me and my colleagues from the factories present our progress on promoting the "Two Methods." From that point on, news of the methods spread throughout the country, and we were soon invited to countless places to share and teach them. During those turbulent ten years, our small team traveled to over 20 provinces, cities, and autonomous regions, visiting thousands of factories, mines, villages, military units, and hospitals. We shared the "Two Methods" with millions of workers, farmers, and soldiers—all in the hope of contributing to the improvement of the national economy.

Creating a supportive environment where scientists can fully apply their expertise is of utmost importance. Premier Zhou understood better than anyone the hearts and hardships of scientific workers and went to great lengths to give them the opportunity to engage with real-world practice. If not for his timely instructions during the "Cultural Revolution," I might not have survived—or, if I had, I would have achieved nothing at all. Any small contribution I was able to make during that time, I owe entirely to Premier Zhou.

Stepping into practice was no smooth sailing. Even after I committed myself to applying mathematical methods in real-world practice, the journey was far from smooth. Some people began to criticize, "Hua Luogeng is only focused on practical work now and no longer cares about theory!" But this, once again, was the same old way of thinking—setting theory and practice at odds with each other. But had I truly separated myself from theoretical research? If that were the case, how could I, after "Cultural Revolution," travel to countries like Britain, France, Germany, and the United States and stand before my peers to both present and seek advice? The fact that I was able to discuss fourteen different scientific issues with international peers was the clearest proof that, while actively promoting the practical application of mathematics, I had never ceased theoretical research. Over the years, I had quietly accumulated a wealth of results still awaiting publication. My ability to sustain this dual focus was inseparable from the consistent guidance of Premier Zhou. A true scientist does not rest on past achievements; what we value is the spirit of innovation and the drive for constant progress—not stagnation, not living off old glories. The results we achieved under the most adverse conditions often astonished foreign peers when we met them, leaving a deep and lasting impression.

The theft of my manuscripts remains unresolved—a matter Premier Zhou had personally addressed in his written instructions. It stands in stark contrast to an incident I encountered during my visit to the California Institute of Technology. There, a distinguished professor showed me two letters I had written to him back in 1939 and 1940, and told me, "You cannot take the originals, but you are welcome to make copies and take them home." In contrast, those who stole my manuscripts—especially the portions directly related to the national economy—should surely feel ashamed. And those who failed to carry out Premier Zhou's instructions, should they not feel even more so? Do they even understand the meaning of respecting knowledge?

The portion of my manuscripts relating to the mathematical theory of large-scale optimization for planned economies—the one "theory" I mentioned earlier—was entirely stolen. But now, not only have I reconstructed it from memory, it has also undergone new development. An abstract of this work, titled "Mathematical Theory of Large-Scale Optimization for Planned Economies," began serial publication in the Chinese *Science Bulletin* starting from issue 12 in 1984. These manuscripts were originally written in the 1960s, stolen in the 1970s, and only recovered and further developed in the 1980s. More than twenty years have passed between their birth and rebirth. The process of reconstruction and development took three full years of my effort—and those were three precious years in the twilight of my life, after I had already passed seventy.

On July 9, 1984, after returning from the United States, I traveled to Changsha, Harbin, Hohhot, and Daqing. Along the way, I saw the promising momentum and bright future of China's development since the Third Plenary Session of the Eleventh Central Committee. Many provinces, cities, and autonomous regions invited me to serve as their Chief Adviser on Science and Technology or Economic Development. But I was deeply aware that, given my knowledge, capability, age, and health, I could not truly live up to such titles. For me, the real work lies in the word "practice"—that is where a lifetime of effort should be spent. I could only interpret these honors as a reflection of the era's respect for knowledge and talent, and as a recognition of the work we had done in the past.

Because of exhaustion and strain, I was admitted to the hospital shortly after returning from Daqing. The doctors advised me to avoid emotional agitation. But when I read the volume II of *Selected Works of Zhou Enlai*, I found it impossible to stay calm. Each page, each speech leaped before my eyes, revealing clear policies, sound strategies, precise goals, concrete measures, and patient, thoughtful guidance—reading them even today feels like a breath of fresh air, illuminating and inspiring. And when I came across Premier Zhou's directives from the "Cultural Revolution," I was overwhelmed with emotion. Those were sharp swords aimed at the "Gang of Four," testimonies that later helped bring them to justice. They are the living embodiment of our beloved Premier Zhou's selfless devotion to protecting scientists, veteran cadres, and the future of our country—a shining example of the noble spirit of sacrifice for the cause of communism, and the heartfelt loyalty of a proletarian revolutionary. We, the countless scientists of China, hold deep admiration for Premier Zhou—for his visionary wisdom and his broad, generous heart.

Dearest Premier Zhou, the sons and daughters of China will cherish your memory forever!

We Must Cultivate a Great Number of People with Genuine Talent and Knowledge*

In China, we must establish a culture that respects knowledge and values talents. We must cultivate and foster a great number of people with genuine talent and knowledge. Why? Respecting knowledge and valuing talents, cultivating people with genuine talent and knowledge, these are not our goals in themselves. Our goal is to achieve the Four Modernizations. However, many people are still not clear about the necessity of cultivating and fostering a great number of people with genuine talent and knowledge to achieve this goal. Genuine talent and knowledge refer to scientific knowledge and abilities, and we must not underestimate their importance. During the construction of the Gezhouba Dam, Zhang Guangdou suggested blasting a gigantic reef, which was adopted. As a result, the dam was well built and withstood the test of floods. If we were to write an article just about blasting the reef, it might not be good or award-winning, but this action was highly effective. When Mr. Sun Yat-sen's household machine broke down, he hired a repairman to fix it, and it was repaired in one minute. The repairman charged fifty yuan and one dime. Mr. Sun asked, "So much money for just one minute?" The repairman replied, "It's just one dime for the minute of repair, but fifty yuan for the knowledge." That's the power of knowledge.

Respecting knowledge naturally leads to respecting talents and actively seeking them out. The Vice President of the University of California, a renowned physicist, repeatedly wrote to Morgan inviting him to teach at the school. Morgan, to decline

* First published in January 1985.

the offer, proposed very demanding conditions. However, to everyone's surprise, the Vice President accepted all of Morgan's conditions. Indeed, Morgan had achievements; during his tenure at the university, he pioneered genetic engineering. The Soviet Union implemented a policy of rewarding scientists and artists who made significant contributions, such as the famous aircraft designer Tupolev, cosmonaut Gagarin, and ballet master Ulanova. America can attract the world's best scientists, showing that their talent policy is effective.

Valuing talents does not equate to valuing diplomas; it means valuing abilities, specifically the practical ability to research and solve problems. Diplomas should only serve as references. I became a professor at National Southwest United University at the age of twenty-eight and a professor in the United States at thirty-eight, but I don't have a PhD, making me the only member of our Academic Department without one. Edison and Faraday also didn't have doctorates. Therefore, we cannot focus solely on diplomas. Our education must emphasize practical results to truly equip students with genuine talent and knowledge, enabling them to be versatile and knowledgeable. Americans evaluate Chinese students as having excellent exam scores but lacking research abilities. Comrade Zhang Guangdou said, "Many engineering science dissertations are third-rate mathematical articles because students lack practical experience and can only use mathematical analysis to fill in the gaps." Students need more freedom to think independently. To achieve versatility and knowledge, we must cultivate flexible thinking. While exams that do not require thought aren't effective for talent selection, what's the use of exams that are off-topic? Another, and even more important, step in improvement is moving from breadth to focus. Mere breadth without focus turns one into nothing more than a dictionary or a library. The shift from broad knowledge to distilled insight is difficult. I advocate a comprehensive selection method that involves long-term thinking to move from vast literature to specific literature.

Therefore, our academic approach should be "broad, specialized, and extensive," which means having a broad foundation, specializing in one's field, and extending one's specialized knowledge to other areas. This requires research; education must be combined with research. In the United States, no professor only teaches without conducting research, nor does anyone only conduct research without teaching. Doctors must also conduct research when treating patients, and those handling administrative tasks must also conduct research. We must also promote the investigation of more problems; its impact on our production, education, and scientific endeavors will be immeasurable.

To cultivate a great number of people with genuine talent and knowledge, we must reform our educational system. Currently, the shortcomings of our educational system include inadequate streamlining, with only one path: elementary school, high school, university, and graduate school, causing overcrowding. The influence of the imperial examination system is evident here. In feudal society, youth rose to prestige in three steps: *xiucai*, *juren*, and *jinshi*.[1] This imperial examination system allowed only a few to succeed, leaving most in miserable conditions. There are examples like Kong Yiji and Fan Jin. When they passed the imperial exam, becoming *juren*—they were so overjoyed that they went mad. Ancient people said, "Passing the imperial exam increases one's worth tenfold." We must establish a new educational system where every industry has prospects and opportunities, aiming for wide prosperity under the examination system. In short, scientific education must be streamlined, practical, and cultivate a great number of people with genuine talent and knowledge.

1 *Xiucai* (literally "cultivated talent") was the lowest degree awarded in the imperial examination system, allowing limited local privileges. *Juren* ("recommended man") was the provincial-level degree, granting broader access to official appointments. *Jinshi* ("presented scholar") was the highest degree, awarded after passing the national examination in the capital, and was a prerequisite for high-ranking government positions.—Trans.

ABOUT THE AUTHOR

Hua Luogeng (1910–1985), mathematician, academician of the Chinese Academy of Sciences, and the founder and pioneer of analytic number theory, is regarded as the "Father of Modern Chinese Mathematics." He made significant contributions to number theory, algebra, and the theory of complex functions. Widely recognized as one of the founders of modern Chinese mathematics, Hua's work not only advanced the field but also played a key role in the development of mathematical education in China.

In addition to his mathematical achievements, Hua was a dedicated educator and advocate for the popularization of mathematics. He worked tirelessly to bring mathematical knowledge to the masses, traveling across China to teach workers and farmers. As a leading figure at the Chinese Academy of Sciences, Hua emphasized the importance of innovation and the cultivation of mathematical talent, shaping the future of Chinese mathematics.

Hua's legacy continues to inspire mathematicians and scholars worldwide, making him a central figure in the history of modern mathematics. His life and work exemplify the transformative power of mathematical education and research.